Laboratory Manual

for
Solomon, Berg, and Martin's

Biology

Seventh Edition

Linda T. Collins

Rebekah P. Bell

Charles H. Nelson

University of Tennessee at Chattanooga

Contributions by Ross M. Durham, Timothy J. Gaudin, & Barbara A. Walton
University of Tennessee at Chattanooga

BROOKS/COLE
CENGAGE Learning

Australia • Brazil • Japan • Korea • Mexico • Singapore • Spain • United Kingdom • United States

BROOKS/COLE
CENGAGE Learning™

Laboratory Manual for Solomon, Berg, and Martin's
Biology, Seventh Edition
Linda T. Collins, Rebekah P. Bell and
Charles H. Nelson

Cover image: "Red-Eyed Tree Frogs" © Chase
Swift/CORBIS

For product information and technology assistance, contact us at
Cengage Learning Customer & Sales Support, 1-800-354-9706

For permission to use material from this text or product,
submit all requests online at **www.cengage.com/permissions**
Further permissions questions can be e-mailed to
permissionrequest@cengage.com

ISBN-13: 978-0-495-01263-4

ISBN-10: 0-495-01263-7

Brooks/Cole Cengage Learning
20 Davis Drive
Belmont, CA 94002-3098
USA

Cengage Learning is a leading provider of customized learning solutions
with office locations around the globe, including Singapore, the
United Kingdom, Australia, Mexico, Brazil, and Japan. Locate your local office at
www.cengage.com/global

Cengage Learning products are represented in Canada by Nelson Education, Ltd.

To learn more about Brooks/Cole, visit **www.cengage.com/brookscole**

Purchase any of our products at your local college store or at our preferred
online store **www.cengagebrain.com**

Printed in the United States of America
4 5 6 15 14

INVESTIGATIONS
IN
BIOLOGY

Rebekah P. Bell
Linda T. Collins
Ross M. Durham
Timothy J. Gaudin
Charles H. Nelson
Barbara A. Walton

The Department of Biological
and Environmental Sciences

The University of Tennessee
at Chattanooga

Chattanooga, Tennessee

TABLE OF CONTENTS

1 Scientific Method.....*1*

2 Enzymes.....*13*

3 Focus on the Cell.....*23*

4 Movement of Molecules in Living Systems.....*39*

5 Fermentation in Yeast - I.....*51*

6 Fermentation in Yeast - II.....*57*

7 Mitosis.....*63*

8 Meiosis.....*77*

9 Heredity.....*93*

10 The Molecular Basis of Heredity I: DNA Structure and Replication.....*111*

11 The Molecular Basis of Heredity II: Protein Synthesis.....*127*

12 Population Biology: The Hardy-Weinberg Equilibrium.....*141*

13 Species Identification and Systematics.....*163*

14 Kingdoms Eubacteria, Protista and Fungi.....*179*

15 Kingdom Plantae: Spore-Bearing Plants.....*197*

16 Kingdom Plantae: Roots, Stems and Leaves.....*209*

17 Kingdom Plantae: Seed Plants.....*223*

18 Kingdom Animalia: Phyla Porifera, Cnidaria and Platyhelminthes.....*245*

19 Kingdom Animalia: Protostomes.....*265*

20 Kingdom Animalia: Deuterostomes.....*281*

21 Development.....*295*

22 Animal Behavior.....*311*

23 Ecology of a Rotting Log.....*323*

Appendix I – Student Safety Contract.....*335*

Appendix II – Laboratory Report Guide.....*337*

EXERCISE 1

SCIENTIFIC METHOD

OBJECTIVES

- Define the vocabulary terms from the questions at the end of this exercise.
- Design an experiment by using the scientific method.
- Distinguish between the experimental group and the control group in an experiment.
- Describe the qualities of a well-designed experiment.

TOPICS

1: OBSERVATION
2: PROBLEM
3: HYPOTHESIS
4: PREDICTION
5: EXPERIMENT
6: CONCLUSION

INTRODUCTION

Defining the term *scientific method* is like defining science. It is impossible to make a definition that satisfies everyone. Most people seem to agree that there is no single method that all scientists follow, but there is a way of thinking that is universal to science. It is a technique of reasoning that first identifies a problem and then attempts to solve it by performing an experiment. It follows a predictable sequence from observation to conclusion. The steps in the scientific method are as follows:

Observation. The starting point is almost always an observation. You might observe something unusual or something very common.

Problem. The next step is recognition of a problem. After making a few preliminary observations, you may have a question about the observation. Most often it takes shape as a "how" or "why" question about the observation.

Hypothesis. A hypothesis attempts to explain an observation. In order to be valid scientifically, it must fulfill several criteria.
- It must suggest a cause-and-effect relationship.
- It must have only one causative agent or variable at a time.
- It must be testable. The variable suggested in the hypothesis must be subject to experimentation.

Prediction. A hypothesis can be used to make a prediction. The prediction states the expected results of the experiment based on the hypothesis. Predictions often take the form of an "if...then..." statement: "If the hypothesis is correct, then the results of the experiment will be..."

Experiment. The function of an experiment is to set up a situation in which the hypothesis can be tested. It is this step that makes science different from pure speculation. An experiment must test only the variable suggested in the hypothesis. This is the independent variable. For example, if your hypothesis is that sowbugs move to the moist, dark, soil end of the container because they prefer moisture, the independent variable is moisture. The dependent variable literally depends on the independent variable. The movement of sowbugs from a moist to dry environment is the dependent variable. A third type of variable is the controlled variable. It is a possible variable that does not change during the experiment. An example of a controlled variable in this experiment is temperature. The temperature does not change and thus does not affect the outcome of the experiment.

Conclusion. If the data support the hypothesis, then the hypothesis is accepted. If the data do not support the hypothesis, then the hypothesis is rejected. As long as an experiment is well designed and executed, the experiment is still valid even when the hypothesis is rejected.

You will use sowbugs as the subjects of your experiment. From your observations you will see a problem or question. For example, you may observe that the sowbugs spend most of their time in one end of the observation chamber. The problem is a question: Why do they spend most of their time in that end of the chamber? Your answer to this problem is the hypothesis. After you state your hypothesis, you will make a prediction based on the hypothesis. Then you will design an experiment to test the hypothesis. You will collect data and reach a conclusion. The conclusion will either support your hypothesis or not.

1: OBSERVATION

<u>PROCEDURE</u>

1. Work in teams of three or four people. Make an observation chamber from an empty box. The chamber should have two distinct areas. One half should have wet soil, and the other half should have dry sand. Be sure that you do not create a barrier to sowbug movement in the middle of the chamber. Put enough soil and sand in the box so the sowbugs can move easily between soil and sand.

2. Place a piece of dark paper on the surface of the soil. Shine a light as closely as possible to the surface of the sand to light and warm the sand. Wait five minutes for the container to achieve equilibrium before continuing.

3. Label the observation chamber in the space provided. Each team member will observe three sowbugs for five minutes each. Team members may start their sowbugs at the same time, but each team member should observe only one sowbug at a time.

4. <u>Your observations</u>. Diagram the movement of the sowbugs. For the first sowbug, place an asterisk * in the drawing of the observation chamber to indicate the starting position of the sowbug. At one-minute intervals for the next five minutes mark the sowbug's location with an X, indicating the time interval for that location. Connect the X's with a broken line to identify the sowbug's movement pattern. After five minutes, remove the first sowbug and replace it with the second sowbug in a different location so the starting replacement is random. Observe the second sowbug for five minutes. Repeat this procedure with the third sowbug so each sowbug has been observed carefully for a five minute period. To avoid confusion, record each sowbug's movement with a different symbol (Y, Z). Keep in mind the primary question as you make your observations. Where do your sowbugs spend most of their time? Record your observations in Table 1.

5. <u>Team observations</u>. When your observations are complete, compare them with those of your team members. Try to determine whether there is any apparent movement pattern that might be common to all the organisms. Record your observations in Table 1. Where do the team sowbugs spend most of their time?

Observation Chamber

Table 1. Sowbug-minutes spent on each side of the observation chamber

	Side 1	Side 2
Observation chamber	Wet, cool, dark soil	Dry, hot, light sand
Team member 1		
Sowbug 1		
Sowbug 2		
Sowbug 3		
Team member 2		
Sowbug 1		
Sowbug 2		
Sowbug 3		
Team member 3		
Sowbug 1		
Sowbug 2		
Sowbug 3		
Team member 4		
Sowbug 1		
Sowbug 2		
Sowbug 3		
Total team sowbug-minutes		
Percent sowbug-minutes		

5. Disassemble your observation chamber. Return the wet soil and dry sand to their containers on the side bench. You will use the empty box for your experiment.

2: PROBLEM

Your observation should give you the basis for posing a question. You have observed one of two possibilities: either the sowbugs preferred one end of the box to the other or they moved randomly. The question is "Why did the bugs have the behavior you observed?" Decide your question as a group, and write it in this space.

3: HYPOTHESIS

Answer the problem by formulating a hypothesis. Consider the set-up you used in making the initial observations and list all the variables that you can identify. You should find at least four. Talk it over with your teammates and decide which variable most influences the observed behavior pattern. The variable that best answers why the sowbugs exhibited the observed behavior is the basis for the hypothesis. Write the hypothesis on the work sheet. Remember that a hypothesis must involve only one variable. State the hypothesis.

4: PREDICTION

Predict the results of your experiment based on your hypothesis using an "if...then..." statement. State your prediction in the space provided.

6

5: EXPERIMENT

Design and perform an experiment with your teammates. Describe your design and state your results in detail so other researchers can repeat your experiment.

<u>PROCEDURE</u>
1. <u>Method.</u>
   Make an experimental chamber by arranging the contents to test your hypothesis. Be sure you are testing only one variable. In the space below, list the variable you are testing and the variables you will control. Check the design with your instructor before you proceed.

2. Label the experimental chamber on the next page.

3. <u>Experiment.</u>
   Test sowbug behavior as you did in the observation. Each team member should test three sowbugs by putting the sowbugs in the test chamber one at a time and watching their movement for five minutes. Count the number of minutes your sowbugs stay in each part of the test chamber. These are sowbug-minutes. Record your sowbug minutes in Table 2.

4. <u>Results.</u>
   Pool the data of all the members of the team to get the number of total team sowbug-minutes for the two parts of the chamber. Record in Table 2. Determine the percent of sowbug-minutes in each part of the chamber. Record your experimental data in Table 2.

5. When you are finished, empty your test chamber back into the proper basins so the materials will be ready for the next lab.

6. Clean up spilled sand and soil on the side bench.

7. Clean your workstation and replace equipment on the tray.

Experimental Chamber

Table 2.   Sowbug-minutes spent on each side of the test chamber

	Side 1	Side 2
Conditions of test chamber		
Team member 1		
Sowbug 1		
Sowbug 2		
Sowbug 3		
Team member 2		
Sowbug 1		
Sowbug 2		
Sowbug 3		
Team member 3		
Sowbug 1		
Sowbug 2		
Sowbug 3		
Team member 4		
Sowbug 1		
Sowbug 2		
Sowbug 3		
Total team sowbug-minutes		
Percent sowbug-minutes		

6: CONCLUSION

Based on the percent of sowbug-minutes spent on each side of the test chamber, the results either support or do not support the hypothesis. State your conclusion.

An experiment where the results do not support the hypothesis is just as valid as one that is supported by results, because it also adds to scientific knowledge. If you have used only one variable and have tested a large sample size, you have performed a good experiment whether the hypothesis is supported or not. If your hypothesis is not supported, you may want to suggest another hypothesis.

VOCABULARY

TERM	DEFINITION
Scientific method	
Problem	
Hypothesis	
Prediction	
Independent variable	
Dependent variable	
Controlled variable	

Data	
Conclusion	

QUESTIONS

1. What is the function of an experiment?

2. How many variables should an experiment have?

3. What is the order of steps in the scientific method?

4. What is a conclusion based on?

12

5. To what does a conclusion always refer?

6. Is an experiment valid if the hypothesis is not supported by the results?

7. How many variables are there in the diagram below?

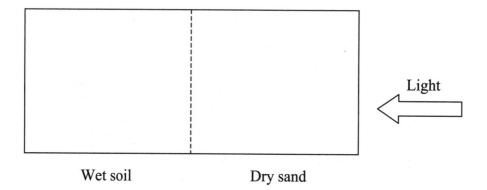

Wet soil   Dry sand

ENZYMES
BIOLOGICAL CATALYSTS

OBJECTIVES
- Define the vocabulary terms from the questions at the end of this exercise.
- Describe the chemical reaction between amylase and starch.
- Explain the effect of temperature on enzyme activity.

TOPICS
1: AMYLASE AT ROOM TEMPERATURE (23°C)
2: AMYLASE AT BODY TEMPERATURE (37°C)
3: AMYLASE IN BOILING WATER
4: AMYLASE IN ICE WATER
5: THE EFFECT OF TEMPERATURE ON REACTION RATE

INTRODUCTION

The chemical reactions inside living cells are controlled by organic catalysts called *enzymes*. In the laboratory, one way to speed up a reaction in a test tube is to apply heat. That is not possible inside a living cell. Without enzymes many reactions inside the cell would be impossibly slow.

Enzymes are catalysts. Catalysts increase the rate of a chemical reaction without being permanently altered themselves. Most enzymes are protein molecules. There are thousands of different enzymes, and each has a very unique three-dimensional shape. Enzymes have an active site, a place with an exact three-dimensional shape where a substrate can fit. The substrate has to fit into the active site of the enzyme in order for the enzyme to catalyze a reaction. Substrates have many different shapes, which explains why enzymes are very specific to their substrates. Excess heat can cause the shape of an enzyme to change so that its active site no longer fits the substrate. If an enzyme can no longer catalyze a reaction, it is denatured.

You will study amylase, an enzyme present in human saliva and the small intestine. Amylase speeds up the hydrolysis of starch into smaller polysaccharides and eventually maltose.

$$\text{Starch} \xrightarrow{\text{amylase}} \text{Polysaccharides and Maltose}$$

The hydrolysis of starch can be monitored by testing with iodine solution. Starch reacts with iodine solution to form a dark blue color. If you test an unknown sample with iodine solution and you get a dark blue color, you know starch is present. If you test an unknown sample with iodine solution and you get a color that is the same as the iodine solution (light yellow), you know starch is not present. You will mix amylase and starch together and test the mixture at intervals with iodine solution to see if starch is present. When a drop of iodine solution no longer produces a dark blue color, the solution no longer contains starch. Endpoint has been reached: the starch is hydrolyzed.

Exercise 2

Recall the Brownian movement you observed in a previous exercise. Brownian movement increases with an increase in temperature. What effect do you think an increase in temperature will have on enzyme action? Will the substrate and enzyme have an increased chance of coming into contact? You will be measuring the rate of hydrolysis of starch at four different temperatures: 0°, 23°, 37° and 100°Celsius. State a hypothesis and a prediction to answer the problem or question "What will happen to the rate of amylase-catalyzed starch hydrolysis when the temperature is increased from 0° to 23° to 37° to 100°Celsius?" Then perform an experiment to test your hypothesis.

PROBLEM What will happen to the rate of amylase-catalyzed starch hydrolysis when the temperature is increased from 0° to 23° to 37° to 100°C?

HYPOTHESIS

PREDICTION

EXPERIMENT

PROCEDURE

1. Work in groups of three or four.

2. Label four test tubes with an "A" for amylase. Label four test tubes with an "S" for starch.

3. Swirl the amylase and starch solutions before pipetting. Pipet 2 mL amylase into each of the four A tubes, and pipet 2 mL starch into each of the four S tubes.

4. You will use one test tube with starch and one with amylase for each of the four temperatures.

1: AMYLASE AT ROOM TEMPERATURE (23°C)

PROCEDURE

1. Work in groups of three or four. Pipet a drop of the starch solution in one well of a spot plate. Using a clean pipet, put a drop of water in another well.

2. Add a drop of iodine solution to each of the two wells and note the colors. The starch and iodine solution is dark blue indicating starch is present. The water and iodine solution is the color of the iodine solution indicating starch is not present. This is the color you will see when the starch is hydrolyzed and endpoint is reached.

3. Pour the contents of one test tube containing amylase into one of the test tubes containing starch. Swirl to mix. This is time zero, immediately before hydrolysis begins.

4. After one minute, remove a drop of the mixture with a clean pipet, put it in a clean well in the spot plate, add a drop of iodine solution and note the color.

5. Continue testing in the same manner at one minute intervals until endpoint is reached.

6. Stop testing when endpoint is reached.

AMYLASE AT ROOM TEMPERATURE

Time	Color
1 minute	
2 minutes	
3 minutes	
4 minutes	
5 minutes	
6 minutes	
7 minutes	
8 minutes	
9 minutes	
10 minutes	
11 minutes	
12 minutes	
13 minutes	
14 minutes	
15 minutes	

2: AMYLASE AT BODY TEMPERATURE (37°C)

PROCEDURE

1. Work in groups of three or four. Fill a large beaker half full of water from the 37°C water bath. Put one test tube containing amylase and one test tube containing starch into the beaker.

2. Place the beaker containing the tubes in the 37°C water bath for ten minutes. Mark your beaker with your team number so you can recover your tubes. You will return the water to the water bath after you have completed this exercise.

3. Take the beaker containing the tubes in the warm water to your table. Pour the amylase tube into the starch tube. Swirl the tube, then replace it in the beaker of warm water at your table.

4. Using the spot plate, test the tubes at thirty second intervals with iodine for the presence of starch. Record your results in the box below.

5. Return the water in the beaker to the 37°C water bath.

6. Stop testing when endpoint is reached.

AMYLASE AT 37°C

Time	Color
0.5 minute	
1 minute	
1.5 minutes	
2 minutes	
2.5 minutes	
3 minutes	
3.5 minutes	
4 minutes	
4.5 minutes	
5 minutes	
5.5 minutes	
6 minutes	
6.5 minutes	
7 minutes	
7.5 minutes	
8 minutes	
8.5 minutes	
9 minutes	
9.5 minutes	
10 minutes	

Exercise 2

3: AMYLASE IN BOILING WATER

You may do the first step of this experiment at the same time as the first step of the ice water experiment.

PROCEDURE
1. Work in groups of three or four.

2. Put a test tube containing amylase in the boiling water bath for ten minutes.

3. Remove the test tube from the boiling water bath and pour it into a starch test tube. Keep the test tubes at your table.

4. Using the spot plate, test for the presence of starch at ten minute intervals. Record your results in the box below.

AMYLASE IN BOILING WATER

Time	Color
10 minutes	
20 minutes	
30 minutes	
40 minutes	
50 minutes	
60 minutes	

4: AMYLASE IN ICE WATER

You may do the first step of this experiment at the same time as the first step of the boiling water bath experiment.

PROCEDURE
1. Work in groups of three or four.

2. Place an amylase test tube and a starch test tube in the ice water on the rear bench for ten minutes.

3. After ten minutes pour the chilled amylase test tube into the chilled starch test tube.

4. Place the starch-amylase tube in the ice water on the rear bench.

5. Obtain a dropperful of your chilled sample from the ice bath on the rear bench at ten minute intervals. Test for the presence of starch and record your results in the box below.

6. Stop testing when end point is reached.

AMYLASE IN ICE WATER

Time	Color
10 minutes	
20 minutes	
30 minutes	
40 minutes	
50 minutes	
60 minutes	

5: THE EFFECT OF TEMPERATURE ON REACTION RATE

PROCEDURE
1. Tabulate the reaction time for each temperature in degrees Celsius. Calculate the reaction rate. Reaction rate is the speed of a reaction. Reaction rate is the reciprocal of time.

	Reaction time minutes to endpoint	Reaction rate 1/minutes to endpoint
0° C		
23° C		
37° C		
100° C		

2. Draw a graph showing the effect of temperature on the reaction rate. Plot the rate of reaction (reciprocal of time) on the Y (vertical) axis. Plot the temperature on the X (horizontal) axis.

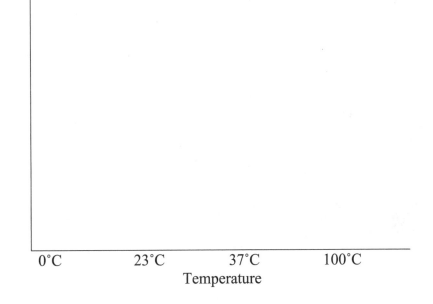

Reaction rate
(1/minutes to
endpoint)

0°C   23°C   37°C   100°C

Temperature

CONCLUSION Is your hypothesis supported?

Exercise 2

20

VOCABULARY

TERM	DEFINITION
Enzyme	
Active site	
Substrate	
Endpoint	
Reaction time	
Reaction rate	
Denature	

Exercise 2

QUESTIONS

1. How do you know when an endpoint is reached after amylase and starch are mixed together?

2. How does the effect of temperature on Brownian movement influence the effect of temperature on enzyme activity?

3. Was your hypothesis supported by your results? If not, how can you explain your results?

4. Review the graphs from this activity.

22

FOCUS ON THE CELL

OBJECTIVES
- Define the vocabulary terms from the questions at the end of this exercise.
- Identify and describe the function of the components of a compound microscope and a stereomicroscope.
- Demonstrate the ability to focus on a specimen using the low and high power objectives of a compound microscope.
- Demonstrate the ability to view larger specimens with a stereomicroscope.
- Observe and describe the differences between plant cells and animal cells.
- Identify representatives of plant, animal, protist and prokaryotic cells: *Elodea*, human cheek cells, protists, bacteria and cyanobacteria.
- List the differences between prokaryotic and eukaryotic cells.

TOPICS
1: THE COMPOUND MICROSCOPE
2: EXAMINATION OF EUKARYOTIC CELLS
   A. PLANT CELLS
   B. ANIMAL CELLS
   C. PROTIST CELLS
3: EXAMINATION OF PROKARYOTIC CELLS
4: THE STEREOMICROSCOPE

INTRODUCTION

The two major groups of cells are prokaryotes and eukaryotes. Prokaryote cells lack nuclei and other membrane-bound organelles. Organisms belonging to Kingdom Prokaryotae are prokaryotes. Bacteria and cyanobacteria are representatives of this kingdom.

Eukaryotic cells are larger and more complex. They contain organelles such as a nucleus, mitochondria, Golgi complex, endoplasmic reticulum and sometimes chloroplasts. The remaining four kingdoms of living organisms, Animalia, Plantae, Protista and Fungi, are eukaryotic. You will use the compound light microscope and the stereomicroscope to study cells.

Study Figure 1 and learn the parts of the compound microscope and their function.

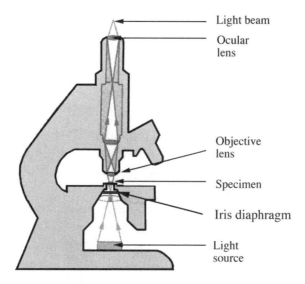

Figure 1. The compound microscope

1: THE COMPOUND MICROSCOPE

The compound microscope is used to study small specimens under high magnification.

Ocular. The ocular or eyepiece is mounted into the top of the microscope body. Compound microscopes may have one or two oculars. The magnification of the ocular is constant and is printed on the side. It is usually 10X, which means it magnifies objects to ten times their actual size.

Objective lenses. The objective lens is found at the bottom of the body tube closest to the slide. In a compound microscope there is more than one objective lens. The lenses are arranged on the revolving nosepiece so that they can be moved into line with the eyepiece. The magnifying power of a microscope is variable: The shorter the lens, the lower the magnification. The total magnification is determined by multiplying the powers of the two lenses you are looking through. If power of the ocular is 10X and the objective lens is 10X, the specimen is magnified 100 times. If the objective lens is 43X the specimen is magnified 430 times.

Focal adjustment knobs. Movement of the coarse adjustment knob causes a relatively large movement of the lens. Use the coarse adjustment knob only with the low power objective. Never adjust downward while looking through the microscope. When you want to adjust downward, you should look at the microscope from the side and turn the coarse adjustment so that the objective moves down very close to the slide you want to view. Then look through the eyepiece and adjust the coarse knob so the objective moves up, away from the slide, to bring the object into view. The fine adjustment is the smaller knob, which moves the lens with more precision. When you want to look at something on high power (43X), first use low power (10X) to find the area you want to view, then

Exercise 3

push the high power objective into place and fine tune the focus with the fine adjustment knob. A microscope is said to be parfocal if only a little fine adjustment is needed to bring a specimen into focus when switching from low to high power.

Iris diaphragm. The iris diaphragm is used to vary the illumination of the object being viewed. On high power you need more light than on low power. However, if there is too much light, the depth of field is decreased, so you should adjust the diaphragm to the lowest light level that permits you to view your specimen easily.

Remember to follow these rules when you use the microscope:
- Always start with the low power objective.
- Never adjust downward while looking through the microscope.
- Use only the fine adjustment knob when using the high power objective.

PROCEDURE
1. Work individually. Choose a clean glass slide from the tray at your table.

2. Place a drop of water on the slide.

3. Obtain a printed slip of paper from the side bench. It will say "objectives."

4. Place the paper in the water; let it get saturated so it will adhere to the slide.

5. Apply a cover slip and blot the bottom of the slide dry. Place the slide on the stage of the microscope.

6. Clip the slide onto the stage with the letters centered over the hole where the light is shining through.

7. Swing the lowest power objective into position over the slide.

8. Looking at the slide from beside the microscope, move the coarse adjustment downward.

9. Looking through the eyepiece, adjust the focus by rolling the coarse adjustment knob toward you until the letters come into view. Use the fine adjustment knob to sharpen the focus. Note the size of the letters on the paper and observe how many you can see in the whole field without moving the slide around.

26

10. Draw the letters as they appear to the unaided eye.

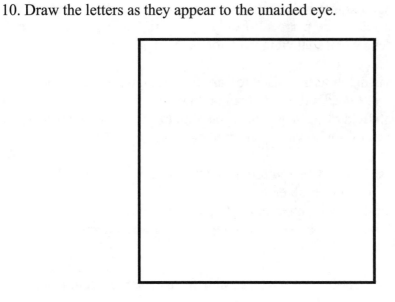

Letters as they appear with the unaided eye

11. Draw the letters as they appear through the microscope.

Letters as they appear with the microscope on low power

12. Move the slide a little to the right. Which way does it seem to go when watching it through the microscope? Pull the slide toward you while watching through the microscope. Which way does it appear to move?

13. While watching the microscope from the side, carefully swing the high power objective over the slide. It will lock into place when it is in the right position. These microscopes are parfocal. Sharpen the focus with the fine adjustment knob. Note the size of the letter now in view. Compare this to the letter size under low power. Draw the letters as they appear with the microscope on high power.

Letters as they appear with the microscope on high power

14. Select another letter such as an "i." Center the dot of the "i" in your field of vision using low power. Switch to high power to see if it is still centered.

15. Switch back to low power to experiment with the amount of light. Flood the field with as bright a light as you can, then focus the fine adjustment up and down slightly to get an understanding of the depth of field. This is the depth at which a 3-dimensional object remains in focus as the adjustment knob is moved. Diminish the light to the lowest level at which you can still see letters. How does the amount of light affect the depth of field? Does more light increase or decrease depth of field?

16. Remove the slide, and then wipe the stage dry with a Kimwipe®. Discard the coverslip in the container for broken glass on the side bench. Rinse the slide and return it to the box or use it for the next exercise.

2: EXAMINATION OF EUKARYOTIC CELLS

A. PLANT CELLS

Plant and animal cells are eukaryotic and have many similarities in internal organization and physiology. However, there are functional and structural differences between them as well. Some of these differences are easily seen using the compound microscope.

Plant cells contain many spherical green chloroplasts where photosynthesis occurs. Cytoplasmic streaming can sometimes be observed as chloroplasts move inside the cells. Cytoplasmic streaming is a movement of cytoplasm that distributes substances within the cell. The central vacuole fills the whole cell interior, but because it is surrounded by the cytoplasm you probably won't be able to see it. Consult your textbook for its location and significance. Plant cells are surrounded by a thick cell wall made of cellulose. The nucleus appears as a gray mass denser than the surrounding cytoplasm. It can often be found just inside the cell wall along the edge of the cell.

PROCEDURE
1. Use a compound microscope and work individually.

2. Place a small *Elodea* leaf from the side bench onto a clean slide and make a wet mount.

3. Using low power, center on an area only a few cells thick near the edge of the leaf and focus.

4. Carefully switch to high power. Sharpen the focus with the fine adjustment.

5. Draw and label the following structures: chloroplasts, central vacuole, cell wall and nucleus.

Plant Cell

6. Explore the depth of field in an *Elodea* leaf. Move the slide until you are looking at a relatively thin part of the leaf near the edge. The image should be clearly visible but should be slightly blurred everywhere else.

7. Using the fine adjustment, raise the lens until the top layer of cells is in focus. Note carefully how much of the cell is clearly visible and not blurred. Can you see the layer immediately below with any clarity?

8. Keep raising the lens until the bottom cell layer passes just out of focus.

9. Raise the lens back to the position it occupied in step 6. Move the high power objective into position. Again note how much of the cell is in focus and how much is blurred. How does the depth of field of low power compare to that of high power? Because the leaf is relatively thick do not lower the high power lens to get the bottom layer in focus. As you can see, we pay a price for an increased ability to magnify. Both the size of the field and the depth of the field decrease as magnification increase.

10. Add a drop of sodium chloride solution, found in a dropper bottle at your table, to the edge of the coverslip. Capillary action will draw the solution under the coverslip. The cytoplasm and cell membrane will pull away from the cell wall as a result of water loss. This phenomenon is called plasmolysis. Next week in lab you will learn why the salt solution causes a water loss inside the cells.

B. ANIMAL CELLS

Cheek cells are part of the tissue type that lines the inside of the mouth. They are characteristically thin, flat and irregular in shape. Sometimes the edge is folded over parts of the cell in a slide preparation. No cell wall is present. If you look very closely you will notice that you cannot see the cell membrane. However, you can detect the point where cytoplasm begins. The interior of the cell is filled with cytoplasm. The large central vacuole that is found in a plant cell is not present in an animal cell. The nucleus is stained dark blue.

PROCEDURE
1. Work individually on this exercise using the compound microscope. You will use a sample of your own cheek cells.

2. Gently scrape the inside of your cheek with a clean toothpick. Force is not necessary to obtain all the cells you will need.

3. Place the cells in a drop of water on a clean slide. Apply a coverslip.

4. Stain the slide preparation with methylene blue dye. Place a small drop of dye along the edge of the coverslip making sure that it contacts the water. Place an absorbent piece of paper towel along the opposite edge of the coverslip. Capillary action will draw the water and dye across the cells under the coverslip.

5. Rinse the excess dye by adding water to one edge of the coverslip and drawing it across the cells by holding a dry piece of paper towel to the opposite edge. Don't let the towel absorb all the water. Some moisture is necessary so the cells don't dry out.

6. Place the slide on the microscope stage and locate the stained cells on low power. When they are focused, switch to high power. Note the differences in appearance between these cheek cells and *Elodea* cells.

7. Draw a cheek cell and label the nucleus, cytoplasm and cell membrane.

Cheek cell

C. PROTIST CELLS

Protist cells are eukaryotic and vary more than any other kingdom in their means of reproduction, nutrition, and locomotion. Most protists are unicellular.

PROCEDURE
1. Work individually. Make several wet mounts from the container of pond water on the side bench. Sample first from the top near floating leaves that may be serving as a food source for other organisms.

2. Next sample from the bottom sludge where there should be many larger, heavier organisms. See if you can find any organisms that appear to be made of more than one cell.

3. Sketch examples of protist cells.

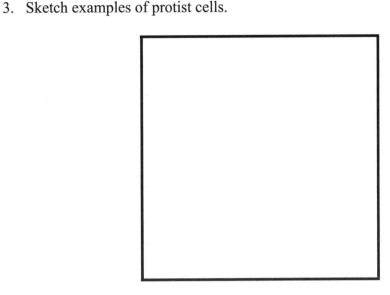

Protist cells

3: EXAMINATION OF PROKARYOTIC CELLS

Prokaryotes lack a nucleus, and their cell walls are different in composition from eukaryote cells walls. Bacteria and cyanobacteria are representatives of this group.

Prokaryotic cells are very small. The highest magnification that can be obtained with your compound microscope is 430X, not enough magnification to see the morphology of most prokaryotic cells clearly. You will observe a cyanobacterium with the compound microscope and look at photomicrographs of bacteria.

PROCEDURE
1. Work individually. Make a wet mount of a cyanobacterium found on the side bench.

2. Using the compound microscope focus first on low power.

3. After focusing the specimen on low power switch to high power and fine focus.

4. Note the difference in size between prokaryotes and eukaryotes. Can you see organelles?

5. Sketch the cyanobacterium.

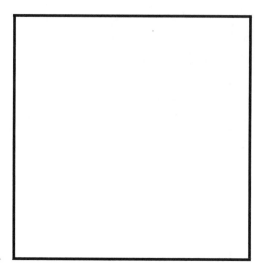

Cyanobacterium

6. Examine photomicrographs of other prokaryotic cells.

4: THE STEREOMICROSCOPE

The stereomicroscope, also called a dissecting microscope, is used to examine larger specimens under low magnification.

Ocular lenses. Stereomicroscopes have two oculars. The distance between them can be adjusted to fit your eyes. The magnification of the oculars is usually 10X.

Objective lenses. The objective lenses vary according to your microscope. Some microscopes have lenses with two different magnifications, and some have zoom capability that allows a continuous increase or decrease of magnification.

Light source. Some stereomicroscopes have two types of illumination. Reflected light illuminates the specimen from above, and transmitted light passes through a translucent specimen from a source below the stage.

PROCEDURE

1. Work individually. On the side bench there are several specimens to view with the stereomicroscope.

2. Observe these specimens with the stereomicroscope and make a sketch of one item.

Specimen viewed through the stereomicroscope

34

VOCABULARY

TERM	DEFINITION
Prokaryote	
Eukaryote	
Parfocal	
Chloroplast	
Cytoplasmic streaming	
Central vacuole	
Cell wall	

Nucleus	
Depth of field	
Plasmolysis	

QUESTIONS

1. Describe briefly the function of the following parts of the light microscope:

   Objective lens

   Fine adjustment knob

   Ocular

   Iris diaphragm

2. What is the total magnification of a microscope using a 10X ocular and a 43X objective?

3. If you move a prepared slide on the stage of a microscope from right to left, what movements do you observe through the ocular?

4. Draw the numeral 4 as it would appear through the microscope.

5. Describe the steps involved in preparing to view a slide on high power.

6. Make drawings of representative plant and animal cells. Label the parts.

Plant cell	Animal cell

7. What type of specimen is best observed with a stereomicroscope?

8. List at least three major differences between plant cells and animal cells.

	Plant cell characteristics	Animal cell characteristics
1.		
2.		
3.		

9. Compare prokaryotic and eukaryotic cells and cite examples of both.

10. Review your drawings from this activity.

EXERCISE 4

MOVEMENT OF MOLECULES IN LIVING SYSTEMS
Brownian Movement, Diffusion, Osmosis, Filtration, Active Transport

OBJECTIVES
- Define the vocabulary terms from the questions at the end of this exercise.
- Observe Brownian movement and understand its significance in biological systems.
- State the physical conditions necessary for diffusion to occur.
- State the physical conditions necessary for osmosis to occur.
- Describe the role of the cell membrane in living systems and consider how it controls the movement of materials.
- Compare the terms isotonic, hypertonic and hypotonic.
- State the physical conditions necessary for filtration to occur and give an example where filtration occurs in the human body.
- State the physical conditions necessary for active transport to occur and give an example of cellular active transport.

TOPICS
1: BROWNIAN MOVEMENT
2: DIFFUSION
3: OSMOSIS
4: FILTRATION
5: ACTIVE TRANSPORT

1: BROWNIAN MOVEMENT

Molecules are in constant random motion due to inherent kinetic energy. They move in a straight line until they bump into another molecule. Then they bounce off in another direction. This motion is named Brownian movement after its discoverer, Robert Brown. Brownian movement is temperature dependent. The higher the temperature, the more rapid the movement will be.

PROCEDURE
1. Work in teams of three or four.

2. Place a small drop of India ink on a slide. Add a drop of water to the India ink. Put a cover glass on the slide.

3. Under high power observe the motion of the ink particles as they bounce off water molecules.

Exercise 4

2: DIFFUSION

Diffusion is the movement of particles from a greater concentration to a lesser concentration until they are equally distributed. Diffusion is a common process in biological systems. For example, gases such as oxygen and carbon dioxide move between blood and cells by diffusion.

PROCEDURE
1. Work in teams of three or four.

2. Obtain a Petri plate filled with agar. Fill the well in the agar with a potassium permanganate solution. Do not let the well overflow. Blot any overflow with a Kimwipe®.

3. Measure in millimeters the movement of the solution from the edge of the well into the agar at fifteen minute intervals for the remainder of the laboratory period. Record your results below and label the time and distance of the growing circles.

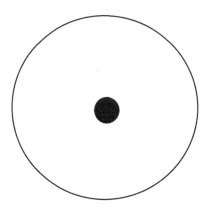

Results from diffusion of potassium permanganate

3: OSMOSIS

Osmosis is the diffusion of water across a selectively permeable membrane. Since osmosis is diffusion of water, the water moves from a region of greater concentration of water to a lower concentration of water. A selectively permeable membrane freely allows the passage of water but stops larger particles. The particles are the solute, and they are dissolved in the solvent, water.

Terms that compare two solutions separated by a selectively permeable membrane are relative terms. For example, you can compare two solutions of table salt dissolved in water. The solutions are separated by a selectively permeable membrane that allows the passage of water but not salt. If the concentration of salt is the same in the two solutions, the solutions are isotonic to each other. Water movement is equal on both sides of the membrane, and net osmosis does not occur. However, if the concentration of salt is greater in one solution than another, the solution with greater salt concentration is hypertonic to the other solution, and net osmosis occurs. Water diffuses because the concentration of water in the hypertonic solution is less than the other solution. If the concentration of salt is less in one solution than the other, the solution with less salt concentration is hypotonic to the other solution. Water diffuses from the solution with greater concentration of water (hypotonic solution) to the solution with less water (hypertonic solution).

The osmotic pressure is the pressure that builds up in a confined solution because of osmosis. The osmotic pressure of a solution is directly proportional to the number of solute molecules or ions in a solution and not the size of the solute particles.

You will place potato cores in various salt solutions ranging from distilled water (0% salt) to 10% salt. The potato cores will increase in weight and length in a hypotonic solution, decrease in weight and length in a hypertonic solution and remain the same in an isotonic solution. After you make up the solutions, ask yourself which salt solution is isotonic to the potato cell.

PROBLEM
What salt solution is isotonic to the potato core?

HYPOTHESIS
State the salt solution you think will be isotonic to the potato core.

PREDICTION
Make a prediction based on your hypothesis.

EXPERIMENT

<u>PROCEDURE</u>

1. Work in teams of three or four. Using the cork borer, cut six potato cores, each about four centimeters long. Do not make the cores longer than four centimeters. Cut all the cores in the potato in the same direction, and work as quickly as possible so they won't dry out. Trim all the cores so they are the same lengths and record their lengths and weights. Do not blot the potato cores or place them on a paper towel. This will cause the potatoes to lose water before you start the experiment.

2. Number six test tubes from 1–6. Perform a standard serial dilution:
   <u>Test tube 1</u>   10% salt
   Measure 20 mL of a 10% salt solution into test tube 1.

   <u>Test tube 2</u>   5% salt
   Pour 10 mL of the solution in test tube 1 into test tube 2. Add 10 mL of distilled water and mix to obtain a 5% salt solution.

   <u>Test tube 3</u>   2.5% salt
   Pour 10 mL of the solution in test tube 2 into test tube 3. Add 10 mL of distilled water and mix to obtain a 2.5% salt solution.

   <u>Test tube 4</u>   1.25% salt
   Pour 10 mL of the solution in test tube 3 into test tube 4. Add 10 mL of distilled water and mix to obtain a 1.25% salt solution.

   <u>Test tube 5</u>   0.625% salt
   Pour 10 mL of the solution in test tube 4 into test tube 5. Add 10 mL of distilled water and mix to obtain a 0.625% salt solution. Discard 10 mL of the 0.625% salt solution.

   <u>Test tube 6</u>   distilled water
   Measure 10 mL of distilled water into test tube 6.

3. Place potato cores in each test tube and let them sit for a minimum of forty-five minutes. Be sure the solutions in the test tubes completely cover the potato cores.

4. Remove the cores from the test tubes. Do not blot the cores.

5. Weigh and measure each core and record your results in Table 1. Graph your results in Graphs 1 and 2.

Exercise 4

Table 1.   Osmosis in potato cores

Test Tube #	Salt Conc.	Before		After		Change	
		Weight (grams)	Length (mm)	Weight (grams)	Length (mm)	Weight (grams)	Length (mm)
1	10%						
2	5%						
3	2.5%						
4	1.25%						
5	0.625%						
6	0%						

Exercise 4

44

Graph 1.   Change in weight of potato cores in salt solutions

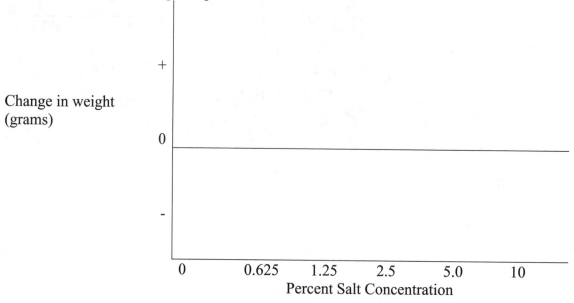

Change in weight (grams)

Graph 2   Change in length of potato cores in salt solutions

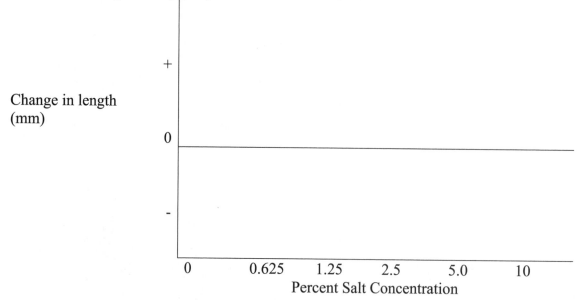

Change in length (mm)

- Which salt solution (if any) is isotonic to the potato?

- Which salt solution(s) is/are hypertonic to the potato?

- Which salt solution(s) is/are hypotonic to the potato?

Exercise 4

CONCLUSION

After evaluating the data, state your conclusion. The results either support or do not support the hypothesis.

4: FILTRATION

Filtration separates molecules based on size. It is a process of pushing small particles and a solvent through a membrane by pressure. Large particles are too big to pass through the membrane. Water, small molecules and ions are separated from larger ones by filtration in the kidney. Blood pressure pushes these small molecules and ions from the capillaries into the collecting system of the kidney. In this experiment pressure is exerted by gravity, and the "membrane" is filter paper.

PROCEDURE

1. Work in teams of three or four. Fold the filter paper as directed and put it in the funnel. Place the funnel in a flask. The flask will catch the filtrate.

2. Using a pipet, measure 5 mL of starch solution into a beaker. Using another pipet, measure 5 mL of 1% copper sulfate solution into the beaker. Add a small scoop of powdered wood charcoal to the beaker. Swirl to mix and pour into the filter paper. Be sure that the solution does not come to the top of the filter paper. Allow the solution to complete filtration.

3. To determine if starch has come through the filter paper, test the filtrate with iodine solution. If starch is present the filtrate will turn dark blue in color. Note the color of the filtrate for the presence of copper sulfate and wood charcoal. Which substances were small enough to pass through the filter paper? Which were too large?

5: ACTIVE TRANSPORT

Fresh water microorganisms live in a hypotonic environment and must have a way to rid themselves of the excess water that is constantly moving in by osmosis. They have an organelle called a contractile vacuole that collects and pumps out the accumulated water by active transport. This activity requires energy, because the water is being pumped against a concentration gradient from a lower concentration to a higher concentration of water. Note the action of *Paramecium* contractile vacuoles on the videotape.

VOCABULARY

TERM	DEFINITION
Brownian movement	
Diffusion	
Osmosis	
Hypertonic	
Hypotonic	
Isotonic	
Selectively permeable membrane	

Exercise 4

Solute	
Solvent	
Filtration	
Active transport	

QUESTIONS

1. Where does diffusion occur in living organisms?

2. Which salt solution(s) were isotonic to the potato? Which salt solution(s) were hypertonic to the potato?

3. Define osmosis.

4. Where in the human body does filtration occur?

5. Explain how substances are separated during filtration.

50

6. Give an example of active transport in a living cell.

7. What kind of activity would you expect to see in the contractile vacuole of a *Paramecium* if you placed it in seawater? Seawater is approximately 4% salt, and a *Paramecium* is approximately 1% salt.

8. In question 7, which is hypotonic, the seawater or the *Paramecium*?

9. Intravenous solutions are isotonic with body cells and fluids. A common intravenous solution is 0.9% saline (salt). What would happen to your red blood cells if you received an intravenous solution of distilled water? Which is hypertonic, the distilled water or the red blood cells?

EXERCISE 5

FERMENTATION IN YEAST - I

OBJECTIVES
- Design a yeast fermentation experiment. You will perform the experiment in the next laboratory exercise.
- Understand the functions of an independent variable, dependent variable, controlled variable and control group.

TOPICS
1: EXPERIMENT DESIGN
2: EXPERIMENT

INTRODUCTION
Fermentation is an anaerobic process where glucose is oxidized via glycolysis to pyruvate. In the absence of oxygen, the pyruvate is converted to ethyl alcohol (C_2H_5OH) and CO_2. The energy that is released powers the generation of ATP from ADP and inorganic phosphate. The chemical equation for fermentation is:

$$C_6H_{12}O_6 \longrightarrow 2\,C_2H_5OH + 2\,CO_2$$
$$\text{glucose} \qquad \text{ethyl alcohol} \qquad \text{carbon dioxide}$$

Since carbon dioxide is a product of fermentation, the amount of fermentation can be determined by measuring carbon dioxide production. In this laboratory exercise you will design a yeast fermentation experiment. In the next laboratory exercise you will perform your experiment.

1: EXPERIMENT DESIGN

What affects the process of yeast fermentation? Possible factors are type of substrate, substrate concentration, enzyme concentration or temperature. When you design your experiment, you will decide which factor you want to test. That factor is the independent variable. For example, if your hypothesis is that an increase in enzyme concentration will cause an increase in rate of yeast fermentation, the independent variable is yeast concentration. The dependent variable literally depends on the independent variable. The amount of carbon dioxide production is the dependent variable. A third type of variable is the controlled variable. It is a possible variable that does not change during the experiment. An example of a controlled variable in this experiment is temperature. Temperature does not change and thus does not affect the outcome of the experiment.

52

<u>PROCEDURE</u>
1. Select an independent variable from the list below.
    ▪   Type of substrate. Yeast will ferment a molecule only if the yeast has enzymes for the necessary reactions. Will yeast ferment these molecules? Glucose, fructose and sorbose are monosaccharides. Sucrose, lactose and maltose are disaccharides. Starch and cellulose are polysaccharides. Saccharine and aspartame are artificial sweeteners.
    ▪   Substrate concentration. If the substrate concentration is increased, will fermentation increase?
    ▪   Yeast concentration. What will happen to fermentation if the yeast concentration is increased?
    ▪   Temperature. What will happen to fermentation when temperature is increased? At what temperature are enzymes denatured?
    ▪   Ethyl alcohol concentration. Ethyl alcohol is a product of fermentation. How much ethyl alcohol can yeast tolerate before ethyl alcohol is toxic?
    ▪   Another variable of your choice.

2. The following materials will be available for you to use. If you want to test a different independent variable, ask your lab instructor if the resources and equipment are available.
    yeast suspension
    10% fructose
    10% glucose
    10% lactose
    10% maltose
    10% sorbose
    10% sucrose
    15 mL and 25 mL test tubes
    5% aspartame
    5% saccharine
    60% ethyl alcohol
    beakers
    graduated cylinders
    plastic pipettes
    water baths at 0°C, 37°C, 45°C

3. Make a list of materials you will need for the next laboratory exercise and give it to your lab instructor.

Exercise 5

2: EXPERIMENT

Make a hypothesis about the effect that your independent variable will have on yeast fermentation. How do you plan to test your hypothesis? What materials will you use?

If you test a variable other than the substrate, use sucrose as the substrate in all test tubes. If you test a variable other than temperature, place your test tubes in a 45° C water bath to increase the fermentation rate. Be sure that you are testing only one variable.

You will measure carbon dioxide production by comparing bubble size at the beginning of the experiment to bubble size at the end of the experiment. Fill a 15-mL test tube with a test solution and yeast. You will invert the test tube in a 25-mL test tube. After incubating the test tubes for one hour, you will measure the change in bubble size in each test tube.

You must have a control group. The control group is different from a controlled variable. The independent variable is usually omitted in the control group. For example, if your independent variable is glucose concentration, use the same amount of yeast that you are using in other test tubes, omit the glucose and fill the test tube with water. If there is any change in bubble size in the test tube by the end of the experiment, you will know that it is not caused by glucose. It is caused by some other factor. If there is an increase in bubble size, subtract the amount from the final results in your other test tubes.

Plan to test 4 differences in your independent variable. For example, if you choose substrate type as your independent variable, test four different substrates. If you choose yeast concentration, test four different yeast concentrations.

Duplicate your results. Fill two test tubes with identical solutions for the control and each change in the independent variable. Average the final results of the two test tubes. Your experiment will be more valid.

PROCEDURE

1. Discuss your experiment with your laboratory instructor.

2. Give your instructor a list of materials that you will need.

3. State your hypothesis and experimental design in the space below.

HYPOTHESIS

54

EXPERIMENT
    MATERIALS

    METHOD

QUESTIONS

1. Define independent variable. Give an example.

2. Define dependent variable. Give an example.

3. Define controlled variable. Give an example. What is the importance of a controlled variable to an experiment?

4.   Define control group. What is the purpose of a control group.

EXERCISE 6

FERMENTATION IN YEAST - II

OBJECTIVES
- Perform a yeast fermentation experiment.
- Collect data and write a laboratory report.

TOPICS
1: HYPOTHESIS
2: EXPERIMENT
    A. MATERIALS AND METHODS
    B. RESULTS
3: CONCLUSION
4: LABORATORY REPORT

1: HYPOTHESIS

PROCEDURE
1. State your hypothesis in the space below.

2: EXPERIMENT

A. MATERIALS AND METHODS

Refer to the previous laboratory exercise. Put 5 ml sucrose in each test tube if you plan to test a variable other than the substrate. Put 5 ml yeast in each test tube if yeast concentration is not a variable. If you are not testing the effect of different temperatures on fermentation, place your test tubes in a 45° C water bath. Do the experiment in duplicate. Fill two test tubes with identical solutions and average the final results. If there is a change in bubble size in the control group, adjust the final results in the test solutions.

PROCEDURE
1. Complete Table 1 by stating the contents of each test tube.

2. Calculate the amount of yeast and test solutions you will need.

3. Stir the yeast suspension. Pour the volume of yeast suspension that you will need into a graduated cylinder. Take it to your lab bench and pour it into a beaker. Using a plastic pipette, measure the appropriate volume of yeast into each test tube.

4. From the stock solution on the side bench, pour the volume of test solutions that you will need into graduated cylinders. Take them to your lab bench. Make dilutions if you are testing substrate concentration. Using a plastic pipette, measure the test solutions into test tubes.

5. The test tubes must be filled to the top before you invert them into the larger tubes as indicated in drawing (a). If necessary, add water to fill the test tubes. Swirl the test tubes to mix the contents.

6. When the small test tube is filled to the top, insert the small test tube into the large test tube as indicated in (b). Insert a pencil to hold the small tube firmly into the large tube. Then turn the test tubes upside down (c).

7. Make a mark on the outside of the large tube with a grease pencil to indicate the air bubble size at the top of the inverted, small test tube. See drawing (c). Label a test tube with a name or initial.

8. Incubate the test tubes in a 45° C water bath if your variable is not temperature.

9. After incubation make another mark at the bottom of the bubble. See drawing (d).

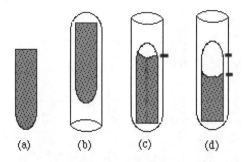

(a)   (b)   (c)   (d)

Table 1. Contents of test tubes

Tube number	Substrate	Substrate ml	Yeast ml	Other variable, if any
1				
2				
3				
4				
5				
6				
7				
8				
9				
10				
11				
12				

B: RESULTS

PROCEDURE

1. Measure the bubble size in the test tubes after one hour. Measure between the grease mark you made at the start of the incubation period and the bottom of the bubble after incubation. Record the measurement in Table 2.

2. The increase in bubble size is caused either by carbon dioxide that is produced by yeast fermentation or by leakage.

60

Table 2. Bubble size (mm)

Tube number	Before Incubation	After Incubation	Change in bubble size	Control change in bubble size	Net change in bubble size	Average change in bubble size
1						
2						
3						
4						
5						
6						
7						
8						
9						
10						
11						
12						

Exercise 6

3: CONCLUSION

PROCEDURE

1. In the space below, state a conclusion based on your hypothesis.

2. If the results do not support your hypothesis, suggest possible sources of error that could have affected the results.

3. Do the results of your experiment suggest other experiments?

4: LABORATORY REPORT

Write a laboratory report to describe your experiment. Include the topics listed below.

PROCEDURE

1. Introduction. State the reason you chose your particular experiment.

2. Hypothesis. State your hypothesis.

3. Materials and Methods. Describe your materials and methods in sufficient detail so the experiment can be repeated.

4. Results. Write a short summary of your results. Be sure you have numerical results. Make a graph of your results.

5. Conclusion. Is your hypothesis supported? Explain your results. Describe other experiments you might want to do.

6. References.

MITOSIS

OBJECTIVES
- Define the vocabulary terms from the questions at the end of this exercise.
- Recognize the stages of mitosis when viewed through the microscope.
- Describe the events of each stage of mitosis.
- Describe cytokinesis in plant cells and animal cells.
- Compare mitosis and cytokinesis in plant cells and animal cells.

TOPICS
1: VIDEO OF MITOSIS
2: PIPE CLEANER REPRESENTATION OF CHROMOSOMES
3: MODELS OF MITOSIS IN ANIMAL CELLS
4: PREPARED SLIDES OF MITOSIS IN PLANT CELLS
5: PREPARED SLIDES OF MITOSIS IN ANIMAL CELLS
6: GIANT SALIVARY CHROMOSOMES OF DROSOPHILA

INTRODUCTION

Mitosis is the process of nuclear division that produces two new nuclei identical to the original nucleus. Cytokinesis is the division of cytoplasm to produce two new cells. In multicellular organisms mitosis and cytokinesis are responsible for growth and maintenance. In unicellular organisms asexual reproduction is accomplished by mitosis and cytokinesis. Mitosis is a continuous process that scientists have divided into stages for the sake of simplicity. The stages merge into one another so smoothly that it is sometimes difficult to determine separate stages. There are several differences between plant and animal mitosis and cytokinesis.

Nondividing cells are in interphase, a stage that is subdivided in order into G_1, S and G_2 (see Figure 1). G_1, the first gap phase, is a period when duplication of cellular organelles begins. Rapidly dividing cells pass through G_1 very quickly and move on to the next stage. G_1 is followed by S, a phase characterized by the replication of DNA. Next comes G_2, the second gap phase, where organelles complete duplication. After the completion of G_2, mitosis begins.

64

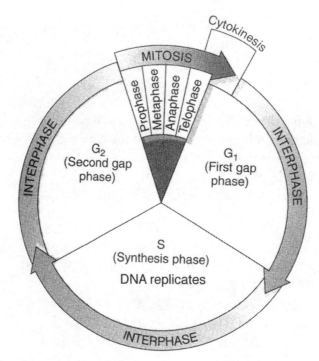

Figure 1. The cell cycle

MITOSIS

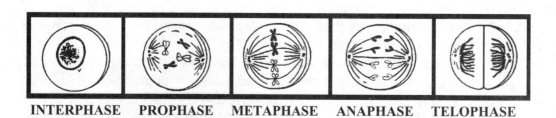

| **INTERPHASE** | **PROPHASE** | **METAPHASE** | **ANAPHASE** | **TELOPHASE** |

Figure 2. Interphase and mitosis

Interphase. A eukaryotic cell in interphase has a clearly visible nucleus and one or more nucleoli. Chromosomes are not visible as rod-like structures. Instead, the nuclear material appears granular. During interphase DNA replicates, making an exact copy of itself, in preparation for nuclear division.

Prophase. During this phase the nuclear membrane and nucleoli disappear and the chromosomes condense. Chromosomes have two identical chromatids, called *sister chromatids*, joined at the centromere region. The spindle becomes visible. In animal cells two centrioles that have replicated during interphase move toward opposite poles of the spindle. Another characteristic of animal cells is the appearance of asters, a series of microtubules that radiate from the centrioles at each pole of the spindle. Centrioles and asters are absent in higher plants. During late prophase in both plant and animal cells spindle microtubules attach to protein plates called *kinetochores* that form on the centromere of each chromatid, thus connecting the centromeres of a chromosome to opposite poles. Other spindle microtubules go from pole to pole.

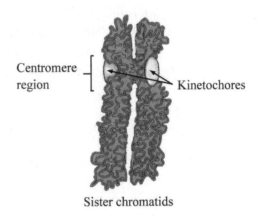

Figure 3. Chromosome

Metaphase. Chromosomes line up on the equatorial plane of the spindle because of changes in the length of microtubules that are attached to the kinetochores. Metaphase ends when the centromeres of each pair of sister chromatids split apart and the chromatids begin to separate.

Anaphase. Sister chromatids separate and move to opposite poles. To accomplish this, the spindle microtubules that are attached to kinetochores shorten, pulling the chromatids apart. Each chromatid is now called a chromosome. The microtubules that stretch from pole to pole lengthen, thus moving the poles away from themselves. By late anaphase each cell contains two groups of widely separated chromosomes.

Telophase. New nuclear membranes form around the two sets of chromosomes. Chromosomes uncoil and the spindle disappears. Cytokinesis, division of the cytoplasm, begins in telophase.

Cytokinesis. Division of the cytoplasm is accomplished by different mechanisms in plant and animal cells. In animal cells a cleavage furrow running the length of the cell pinches the cytoplasm into two portions with a set of chromosomes in each portion. The location of the cleavage furrow is determined by the orientation of the mitotic spindle, with the furrow forming at the equatorial plane of the spindle. In plant cells the rigid cell wall prevents the development of a furrow. A cell plate forms halfway between the two sets of chromosomes. Cytokinesis is completed and the new daughter cells are in interphase.

Exercise 7

1: VIDEO OF MITOSIS

You will see a videotape of mitosis. Look for the stages of mitosis and the differences between plant and animal cells. A time-lapsed sequence will demonstrate that mitosis is a continuous process.

2: PIPE CLEANER REPRESENTATION OF CHROMOSOMES

PROCEDURE
1. Work in teams of three or four.

2. G_1 interphase. To simulate G_1 interphase, take one of each of the following four pipe cleaners: green long, blue long, green short and blue short. Put them on the table. They represent the four chromosomes in a cell. The blue pipe cleaners represent paternal chromosomes and the green pipe cleaners represent maternal chromosomes.

3. S interphase. Pick up the long green pipe cleaner and twist it one time with another long green pipe cleaner to form a pipe cleaner pair that represents a chromosome composed of two chromatids held together at the centromere (twist). Repeat this with the other pipe cleaners so you have one long green chromosome, one long blue chromosome, and one short green chromosome and one short blue chromosome. This replication occurs in S interphase.

4. G_2 interphase. G_2 interphase represents the time between chromosome replication and the actual start of mitosis that begins with prophase.

5. Prophase. Place the pipe cleaners on the table in any order to represent prophase.

(green pipe cleaners)

(blue pipe cleaners)

(green pipe cleaners)

(blue pipe cleaners)

6. Metaphase. Line the twisted pipe cleaner pairs up in a straight line on the equatorial plane. Their order does not matter. This represents the lining up of chromosomes on the equatorial plane in metaphase.

7. Anaphase. Now untwist the pipe cleaner pairs and move the pipe cleaners apart. This represents the separation of sister chromatids and their migration toward the poles of the spindle in anaphase.

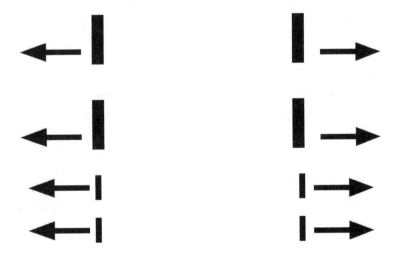

8. Telophase. You now have two cells with four chromosomes each. The two cells are identical to the original cell.

3: MODELS OF MITOSIS IN ANIMAL CELLS

Models are arranged along the side bench. These clearly illustrate the stages you are expected to learn. Make sketches in the boxes in Figure 4 for further study.

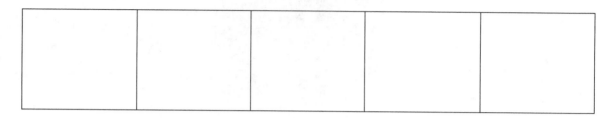

Figure 4. Drawings of mitosis from models

4: PREPARED SLIDES OF MITOSIS IN PLANT CELLS

Root tips are chosen to study cell division because they grow rapidly and provide many cells in the process of mitosis and cytokinesis.

PROCEDURE
1. Work individually. Obtain a prepared slide of onion root-tip (*Allium*) from the side bench.

2. Using low power on the microscope, find all mitotic stages. Then switch to high power and study each stage carefully. Note the absence of asters. Telophase begins when a cell plate begins to form. Make sketches in the boxes in Figure 5 for further study.

Figure 5. Drawing of mitosis in plant cells (*Allium* prepared slides)

5: PREPARED SLIDES OF MITOSIS IN ANIMAL CELLS

The prepared slides that you will study as representative of animal cells are from the whitefish blastula. A blastula is an early stage in the development of an animal embryo. The cells are large and the rate of division is high.

PROCEDURE
1. Work individually. Obtain a prepared slide of the whitefish blastula from the side bench.

2. Using low power on the microscope, find all mitotic stages. Then switch to high power and study each stage carefully. Note the differences between animal mitosis and the plant mitosis. The cells are not oriented in a uniform fashion. Some mitotic cells will be seen cut obliquely or transversely. Concentrate on those that can be seen in full view and are easily identified. Telophase begins when a cleavage furrow begins to form. Make sketches in the boxes in Figure 6 for further study.

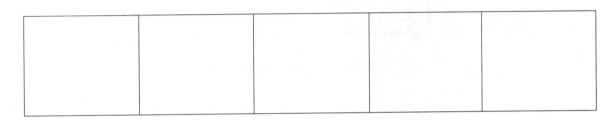

Figure 6. Drawings of mitosis in animal cells (Whitefish blastula prepared slides)

6: GIANT SALIVARY CHROMOSOMES OF DROSOPHILA

Salivary gland cells of the common fruit fly *Drosophila melanogaster* contain giant chromosomes, which are composed of many copies of daughter chromosomes that have been replicated many times but have not separated. The large size of these chromosomes has been valuable in genetic work since banding can be seen. This banding is thought to represent the location of genes. Each band contains one gene copied approximately one thousand times and then stacked side by side. Observe that the thickness of the chromosomes varies from one area to the next and that the width of the bands varies also. Such structural landmarks are specific. The bands allow researchers to pinpoint precisely the chromosome location that may have been affected by experimental treatment of fruit flies.

70

PROCEDURE

1. Work individually. Observe the photomicrograph of *Drosophila* salivary cell chromosomes on the side bench.

2. Draw a chromosome in the space provided below.

Drosophila salivary chromosomes

VOCABULARY

TERMS	DEFINITIONS
Mitosis	
Cytokinesis	
Nucleus	
Nucleolus	
Interphase	
G_1 phase	

G₂ phase	
S phase	
Prophase	
Metaphase	
Anaphase	
Telophase	

Spindle	
Microtubule	
Kinetochore	
Chromosome	
Centromere	
Centrioles	

74

Asters	
Cleavage furrow	
Cell plate	

QUESTIONS

1. Contrast mitosis and cytokinesis in plant cells and animal cells.

Plant	Animal

2. During which phases would you best be able to see mitotic structures that distinguish plant cells from animal cells?

3. When does DNA replication occur in preparation for mitosis?

4. Review your drawings from this activity.

EXERCISE 8

MEIOSIS

OBJECTIVES
- Define the vocabulary terms from the questions at the end of this exercise.
- Recognize the stages of meiosis when viewed through the microscope.
- Describe the events of each stage of meiosis.
- Discuss the significance of the pairing of homologous chromosomes.
- Understand why random assortment of homologous chromosomes is important to genetics.
- Compare and contrast mitosis and meiosis.
- Recognize the differences between oogenesis and spermatogenesis.

TOPICS
1: VIDEO OF MEIOSIS
2: PIPE CLEANER MODEL OF MEIOSIS
3: RANDOM ASSORTMENT OF CHROMOSOMES
4: MEIOSIS IN PLANTS
5: MEIOSIS IN ANIMALS
   A. SPERMATOGENESIS
   B. OOGENESIS

INTRODUCTION
Mitosis produces two cells identical to the original cell. In contrast, meiosis produces four cells that have half the number of chromosomes of the original cells. Meiosis is required in sexually reproducing organisms because gametes must have half the number of chromosomes. When gametes fuse, the diploid chromosome number is restored. Meiosis reduces the chromosome number from diploid (2n) to haploid (n).

Diploid cells have chromosomes in pairs called homologous chromosomes. They bear genes for the same traits. One of the pair came from the male parent and the other came from the female parent. For example, a human somatic cell has forty-six chromosomes, twenty-three pairs of homologous chromosomes. Gametes have twenty-three chromosomes, one from each homologous pair. Homologous pairs line up randomly before they move apart, so the resulting cells have a random mixture of paternal and maternal chromosomes. This mixture allows a great increase in the genetic variability of offspring.

Before a cell undergoes meiosis, DNA replication occurs in interphase. The result is that each chromosome is composed of two identical chromatids. Meiosis has two distinct parts — Meiosis I where the chromosome number is halved and Meiosis II where chromatids separate.

MEIOSIS

Figure 1. Meiosis

MEIOSIS I

Prophase I. Chromosomes condense and become visible. The nuclear membrane breaks down and nucleoli disappear. Homologous chromosomes move together and lie side by side to form a tetrad composed of four chromatids. These four chromatids can exchange fragments by a process called *crossing over*. The spindle microtubules begin to form.

Metaphase I. The tetrads line up singly on the equatorial plane. A kinetochore microtubule from the centromere of one homologue is attached to one spindle pole, and the centromere of the other homologue is attached by a kinetochore microtubule to the opposite pole.

Anaphase I. Homologous chromosomes are pulled to opposite poles as kinetochore microtubules shorten. The centromeres do not divide as they do in mitosis, so each chromosome still consists of two joined chromatids.

Telophase I. Cytokinesis occurs resulting in two daughter cells that are haploid. Nuclear membranes form around chromosomes that have begun to uncoil. Telophase I can be followed by a brief period called *interkinesis*.

MEIOSIS II

Meiosis II is similar to mitosis. Both cells produced by meiosis I will divide again in the meiosis II, but for simplicity we will follow only one cell.

Prophase II. The nuclear membrane breaks down and chromosomes condense.

Metaphase II. The chromosomes line up at the equatorial plane of the spindle. Each chromosome sends kinetochore microtubules in both directions to opposite poles. At the end of metaphase II each centromere splits.

Anaphase II. Sister chromatids separate and move toward opposite poles of the spindle.

Telophase II. Cytokinesis occurs and nuclear membranes form around chromosomes, which have begun to uncoil.

1: VIDEO OF MEIOSIS

2: PIPE CLEANER MODEL OF MEIOSIS

Students should do the following exercise carefully. Pipe cleaners will be used to represent meiosis in the same way they were used in Exercise 5. Our hypothetical organism has two pairs of chromosomes giving a diploid number of four (2n = 4). Each pipe cleaner will represent a chromosome. The blue pipe cleaners represent paternal chromosomes and the green pipe cleaners represent maternal chromosomes.

PROCEDURE
1. Work in teams of three or four.

2. To simulate interphase and DNA replication, take one of each of the following four pipe cleaners: green long, blue long, green short and blue short. Put them on the table. They represent the four chromosomes in a cell. Pick up the long green pipe cleaner and twist it with another long green pipe cleaner with a single twist to represent a chromosome composed of two chromatids held together at the centromere (twist). Repeat this with the other pipe cleaners so you have one long green pipe cleaner pair, one long blue pipe cleaner pair, one short green pipe cleaner pair and one short blue pipe cleaner pair. As in mitosis, this represents the replication of DNA occurring in interphase.

MEIOSIS I

3. Prophase I. Lay the long green pipe cleaner pair and the long blue pipe cleaner pair together on the table so they overlap. The long pipe cleaner pairs are homologues. They bear genes for the same traits. The blue pipe cleaner pair came from the individual's father, the green pipe cleaner pair from the mother. The overlapping of the pipe cleaners represents crossing over. The four chromatids that make a homologous pair are the tetrad. Bring the short pipe cleaner pairs together in the same manner. They too are homologues.

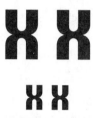

4. Metaphase I. Line the pipe cleaner pairs up on the equatorial plane so that the long pipe cleaner pairs are above the short pipe cleaner pairs.

5. Anaphase I. Pull the pipe cleaner pairs apart in the following manner. This represents the movement of the homologous chromosomes to opposite poles.

6. Telophase I. There are now two cells. Each cell has two pipe cleaner pairs, half the number you started with. Thus, the cells are haploid.

MEIOSIS II

7. Prophase II. The two cells each have one large pipe cleaner pair and one small pipe cleaner pair. It does not matter if each cell has all blue, all green or one of each color.

8. Metaphase II. As in mitosis, line the pipe cleaner pairs up in a straight line on the equatorial plane in each cell.

9. Anaphase II. As in mitosis, untwist the pipe cleaners and move them apart. This represents the separation and movement of the sister chromatids toward the poles.

10. Telophase II. Now there are four cells, each with a haploid number of pipe cleaners. Each cell has one homologue from each original pair, one long pipe cleaner and one short pipe cleaner.

3: RANDOM ASSORTMENT OF CHROMOSOMES

Chromosome combinations differ from one gamete to another. Consider a cell with two pairs of chromosomes. The homologous chromosomes line up on the equatorial plane during metaphase I of meiosis in two possible ways. Both the chromosomes of maternal origin can align on one side of the equator and both the chromosomes of paternal origin on the other side, or each side can have one maternal and one paternal chromosome.

PROCEDURE

1. Work in teams of three or four. Use the pipe cleaner pairs from the previous activity. You should have a long green chromosome, a short green chromosome, a long blue chromosome and a short blue chromosome.

2. Line up the homologous chromosomes as they would be arranged during metaphase I.

3. List the possible combinations of chromosomes for gametes.

4. The number of possible combinations can be determined mathematically. There are only two choices: the chromosome will be on the left side of the equator or on the right side. The number of choices, two, is raised to the power of the number of homologous chromosome pairs. An organism with four chromosomes (two pairs) has $2^2 = 4$ different possible gametes.

$$2^n = \text{possible gametes}$$
$$n = \text{number of homologous pairs}$$

5. In humans there are twenty-three pairs of chromosomes. Calculate the number of different chromosome combinations in human gametes.

Exercise 8

4: MEIOSIS IN PLANTS

In lilies, meiosis occurs in separate parts of the flower to produce haploid male and female gametes. Meiosis occurs in the anther to produce pollen, the precursor of the male gamete. Meiosis occurs in the ovary to produce haploid cells that give rise to the female gamete. During pollination a haploid male gamete from a pollen grain unites with a haploid female gamete embedded in the ovary of the flower to form a diploid zygote. The zygote divides by mitosis and develops into a plant embryo surrounded by nutrients and a protective seed coat.

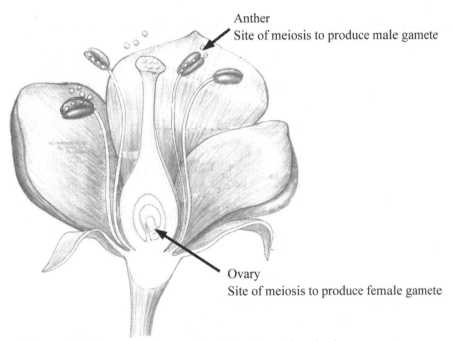

Anther
Site of meiosis to produce male gamete

Ovary
Site of meiosis to produce female gamete

Figure 2. Flower diagram showing sites of meiosis

FORMATION OF THE POLLEN GRAIN

PROCEDURE
1. Work individually.

2. Using the microscope, examine the prepared slides of a cross section of a flower. Both male and female parts will be visible. Find the anther cross sections located in a ring around the perimeter of the specimen. Meiosis occurs in cells inside the pollen sacs of the anther to produce haploid pollen.

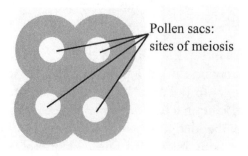

Anther cross section

Early prophase. The chromosomes condense. Nuclei are diploid.

Early prophase I
Diploid cell inside pollen sac

Late prophase. Homologous chromosomes pair.

Late prophase I
Diploid cell inside pollen sac

First meiotic division. Metaphase, anaphase and telophase stages may be visible in the cells of the pollen sacs. Meiosis I is characterized by segregation (separation) of homologous chromosomes. You will see the haploid products of the first meiotic division. The chromosomes of these cells have two chromatids.

Haploid products of meiosis I

Second meiotic division. Groups of two haploid cells are seen in the pollen sacs. Cells are in late prophase, metaphase or anaphase of the second division may be seen.

Tetrads. Tetrads, or clusters, of four haploid cells that are the result of the second division of meiosis are seen. Each of these four haploid cells will develop into a pollen grain.

Haploid pollen tetrad

3. Sketch each stage in the space provided.

	Prophase	Metaphase	Anaphase	Telophase
Meiosis I				
Meiosis II				

5: MEIOSIS IN ANIMALS

Like plants, sexual reproduction in animals requires the formation of haploid gametes, the sperm and egg (ovum). In human males, spermatogenesis, the formation of sperm, occurs in the testes. Spermatogenesis begins at puberty and continues throughout adulthood. Over 400,000,000 functional sperm are produced each day. In human females, oogenesis, the formation of ova, begins in the ovary before birth. At puberty, a partially mature ovum is released an average of once per menstrual cycle until menopause. See Figure 3 for a summary of the differences between oogenesis and spermatogenesis.

A. SPERMATOGENESIS: FORMATION OF SPERM

Spermatogenesis is the formation of haploid sperm. Primary spermatocytes are diploid cells that undergo meiosis. The first meiotic division produces two haploid cells called *secondary spermatocytes*. The result of the second meiotic division is four spermatids. Differentiation occurs to form mature spermatozoa.

PROCEDURE

1. Work individually. Study Figure 3 to see the cells produced by meiosis I and meiosis II during spermatogenesis.

2. Observe photomicrographs of human spermatogenesis found on the side bench. Draw mature spermatozoa in the space provided.

Mature spermatozoa

B. OOGENESIS: FORMATION OF OVA

Oogenesis results in the formation of haploid ova. Primary oocytes are diploid cells that undergo meiosis. During oogenesis the spindle forms on the edge of the primary oocyte rather than in the center, causing unequal cytokinesis. The products of meiosis I are a secondary oocyte and a small, nonviable cell called a *polar body*. The secondary oocyte undergoes the second meiotic division to produce an ovum and another polar body. The polar body produced by the first meiotic division usually does not complete the second meiotic division. Thus the final products of oogenesis are one large ovum and a smaller polar body.

In the ovary the primary oocyte is surrounded by a capsule-like structure called a *follicle*. A follicle matures each month and an immature ovum that is arrested in metaphase of meiosis II is released. Fertilization by a sperm triggers the completion of meiosis. The haploid nuclei from the ovum and sperm fuse to form a zygote, and the normal diploid chromosome number is restored.

PROCEDURE

1. Work individually. Study Figure 3 to see the cells produced by meiosis I and meiosis II during oogenesis.

2. Observe photomicrographs of a human ovary found on the side bench. Draw a follicle containing a secondary oocyte in the space below.

Follicle with secondary oocyte

Exercise 8

OOGENESIS AND SPERMATOGENESIS

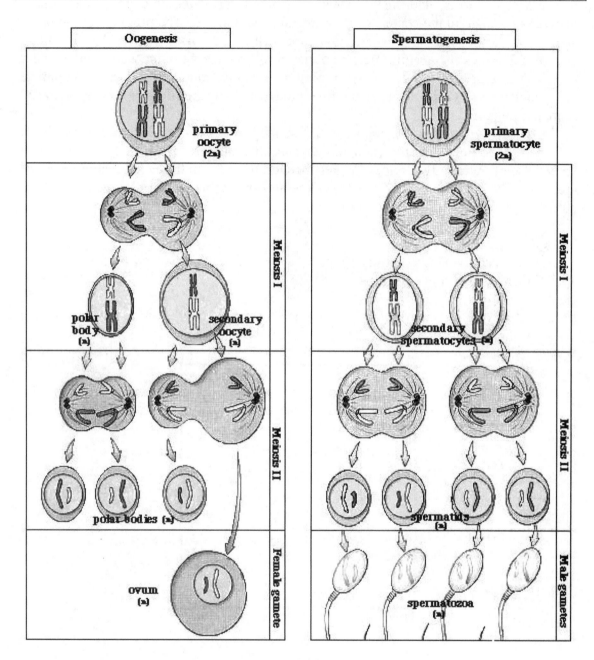

Figure 3. Spermatogenesis and Oogenesis

Exercise 8

VOCABULARY

TERMS	DEFINITIONS
Haploid	
Diploid	
Genetic variability	
Homologous chromosomes	
Tetrad	
Crossing over	
Anther	
Spermatogenesis	
Primary Spermatocyte	

90

Secondary spermatocyte	
Spermatid	
Spermatozoa	
Oogenesis	
Primary oocyte	
Secondary oocyte	
Polar body	
Ovum	
Fertilization	

Exercise 8

QUESTIONS

1. In humans the diploid number is forty-six. How many chromosomes would you expect to find in cells at the following stages?

   _____ spermatid

   _____ primary oocyte

   _____ first polar body

   _____ ovum

   _____ secondary spermatocyte

2. A diploid organism has eight chromosomes. How many different combinations of chromosomes are possible in the gametes?

3. Compare the products of oogenesis and spermatogenesis.

4. Why is it important that sperm and ova are haploid?

5. Summarize in your own words the process of meiosis.

6. Review your drawings from this activity.

EXERCISE 9

HEREDITY

OBJECTIVES

- Define the vocabulary terms from the questions at the end of this exercise.
- Use a Punnett square to determine genotypic and phenotypic ratios expected from any given cross.
- State Mendel's first and second laws of inheritance.
- Observe examples of Mendel's monohybrid and dihybrid crosses and perform a chi-square test on his experimental results.
- Differentiate between a monohybrid cross and a dihybrid cross.
- State the function of a test cross and be able to recognize a test cross.
- Formulate a hypothesis for the genotype of an ear of Indian corn, test the hypothesis and perform a chi-square test on the experimental results.

TOPICS

1: MONOHYBRID CROSS AND CHI-SQUARE ANALYSIS IN PEA PLANTS
2: DIHYBRID CROSS AND CHI-SQUARE ANALYSIS IN PEA PLANTS
3: THE TESTCROSS
4: MONOHYBRID CROSS OF INDIAN CORN
5: DIHYBRID CROSS OF INDIAN CORN
6: DETERMINATION OF AN UNKNOWN INDIAN CORN GENOTYPE

INTRODUCTION

In diploid cells, chromosomes occur in homologous pairs that separate during meiosis so that gametes receive only one chromosome of the pair. Genes for a given trait ordinarily occur at the same location on each homologous chromosome. These genes may have alternate forms called alleles. One allele may be dominant over the other allele. When an allele is present and is always expressed, it is a dominant allele. An allele that is not always expressed is a recessive allele. An individual that has two identical alleles is homozygous. If both alleles are dominant the individual is homozygous dominant. If both alleles are recessive the individual is homozygous recessive and the recessive trait will be expressed. If the alleles are different, the individual is heterozygous, and if one allele is dominant over the other, the dominant one will be expressed.

Phenotype is a term used to describe an individual's appearance as opposed to genotype, or genetic make-up. Phenotype comes from a Greek word that means "to appear." Phenotype does not always reveal the genotype. The genotype of an individual that expresses a recessive trait must be homozygous recessive. The genotype of an individual that expresses the dominant trait could be homozygous dominant or heterozygous. If an individual expressing a dominant trait is crossed with another individual with that dominant trait and there are any offspring with the recessive trait, then both individuals must be heterozygous. In a testcross an individual expressing a dominant trait is crossed with another individual expressing the recessive trait. If any offspring have the recessive trait, then the individual expressing the dominant trait is heterozygous.

1: MONOHYBRID CROSS AND CHI-SQUARE ANALYSIS IN PEA PLANTS

A monohybrid cross examines only one trait at a time. Gregor Mendel performed thousands of monohybrid crosses using the common garden pea plant. One of the traits Mendel studied was flower color. He crossed red-flowered plants with white-flowered plants. This was the P (parent) generation. He observed that the white-flowered plants disappeared in the second generation, the F_1 (filial) generation, and reappeared as approximately one-fourth of the offspring in the third generation, or F_2 generation, when the F_1 generation was crossed with itself. Mendel used the scientific method in his experiments. He formulated a hypothesis to explain his observations. The hypothesis was the basis for his first law, the law of segregation.

Observation. When Mendel crossed red-flowered pea plants with white-flowered pea plants, the F_1 generation was all red-flowered. When plants from the F_1 generation were crossed with each other, the white-flowered plants reappeared as 1/4 of the F_2 generation.

Problem. What can explain the observation?

Hypothesis. Mendel's hypothesis was the basis for his first law, the law of segregation. Each pea plant carries pairs of alleles for flower color. The P generation red plants have two red alleles and the white plants have two white alleles. The alleles segregate when gametes are formed so offspring receive one allele from each parent. The allele for red flowers (A) is dominant over the allele for white flowers (a). Therefore, the F_1 generation plants will all have red flowers. Crossing the F_1 generation with itself will yield offspring that are 3/4 red-flowered and 1/4 white-flowered. See Tables 1 and 2.

Table 1.   Monohybrid cross.
P generation cross to yield the F_1 generation. *A* is the allele for red flowers and *a* is the allele for white flowers. The red-flowered plant produces two *A* alleles. The white-flowered plant produces two *a* alleles.

	A	*A*
a	*Aa*	*Aa*
a	*Aa*	*Aa*

Exercise 9

Table 2.   Monohybrid cross.
        F_1 x F_1 to yield the F_2 generation. The red-flowered plant produces one A allele and one a allele.

	A	a
A	AA	Aa
a	Aa	aa

Prediction. If a red flowered pea plant is crossed with a white flowered pea plant, then all offspring will be red flowered. If these red flowered pea plants are self-crossed, then 3/4 of the offspring will be red flowered and 1/4 will be white flowered.

Experiment.
   P generation: Cross red-flowered plants with white-flowered plants.
   F_1 generation: All plants have red flowers. Cross the F_1 generation with itself.
   F_2 generation: 705 plants have red flowers and 224 have white flowers.

Conclusion. Mendel's hypothesis states that the ratio of plants in the F_2 generation will be 3/4 red-flowered to 1/4 white-flowered. Are his results close enough to the ones expected by his hypothesis? Can you accept his hypothesis? Of the total number of plants (705 + 224 = 929), 3/4 should be red-flowered. Thus, based on the hypothesis, the number expected to be red is 697 (929 x 3/4). The number expected to be white is 232 (929 x 1/4). Mendel observed 705 red and 224 white flowered plants. Is the difference between his observed results and the expected results due to chance or is his hypothesis incorrect? To determine if the observed results are close enough to the expected results, we must submit them to a chi-square analysis. Chi-square is a statistical test that can be used to determine whether results are close enough to those expected by a hypothesis. Chi-square tells you how many times out of 100 a deviation from the expected results is due to chance alone. It is the probability (expressed in percent) that chance alone has caused the deviation from the expected results. If chance has caused the difference between observed and expected results, then the results support the hypothesis.

Chi-square (X^2) is calculated using the following formula:

$$X^2 = \sum \frac{(O - E)^2}{E}$$

Chi-square is the sum of the observed (O) data minus expected (E) data squared, $(O - E)^2$, divided by the expected data in all groups.

$$X^2 = \frac{(705 - 697)^2}{697} + \frac{(224 - 232)^2}{232}$$

$$X^2 = \frac{(8)^2}{697} + \frac{(-8)^2}{232}$$

$$X^2 = \frac{64}{697} + \frac{64}{232}$$

$$X^2 = 0.37$$

These calculations are summarized in Table 3.

Table 3. Chi-square analysis for Mendel's cross between red-flowered pea plants and white-flowered pea plants

Phenotype	O	E	(O-E)	$(O-E)^2$	$\frac{(O-E)^2}{E}$
Red flowers	705	697	8	64	0.09
White flowers	224	232	-8	64	0.28
$X^2 = 0.09 + 0.28 = 0.37$					
$p = 0.50 - 0.70$					

What does the chi-square value signify? To interpret the chi-square value:

- Find the degrees of freedom. Degrees of freedom are equal to the number of groups minus one. In this example, there are two groups (red flowers and white flowers), so there is one degree of freedom.

- Look at the chi-square table (Table 4). Find the closest probability (p) value associated with your chi-square and degrees of freedom. The degree of freedom is one. The chi-square value is 0.37. Therefore, the probability lies between 0.50 and 0.70. Probability is a percentage. There is a 50% to 70% probability that the deviation from the expected results is due to chance alone.

- The criterion for accepting or rejecting the hypothesis is $p > 0.05$. If the probability or p value is greater than 0.05, the hypothesis is accepted. Since the p value of Mendel's experiment is between 0.50 and 0.70, his hypothesis is accepted.

Table 4. Table of chi-square values

Degrees of freedom	PROBABILITY									
	0.95	0.90	0.80	0.70	0.50	0.30	0.20	0.10	0.05	0.01
1	0.004	0.02	0.06	0.15	0.46	1.07	1.64	2.71	3.84	6.64
2	1.10	0.21	0.45	0.71	1.39	2.41	3.22	4.60	5.99	9.21
3	0.35	0.58	1.01	1.42	2.37	3.66	4.64	6.25	7.82	11.34
4	0.71	1.06	1.65	2.20	3.36	4.88	5.99	7.78	9.49	13.28
Hypothesis	Accept								Reject	

98

2: DIHYBRID CROSS AND CHI-SQUARE ANALYSIS IN PEA PLANTS

A dihybrid cross is a cross that examines two traits at a time. Two of the traits that Mendel studied were seed color and texture. He crossed pea plants that had round yellow seeds with pea plants that had wrinkled green seeds. This was the P (parent) generation. The wrinkled green seeds disappeared in the F_1 generation and reappeared in the F_2 generation. The F_2 generation had four phenotypes in the following proportion: nine round yellow seeds; three round green seeds; three wrinkled yellow seeds; and one wrinkled green seeds. Mendel formulated a hypothesis to explain his observations. The hypothesis was the basis for his second law, the law of independent assortment.

Observation. When Mendel crossed round yellow-seeded pea plants with wrinkled green-seeded plants, the wrinkled green traits were lost in the F_1 generation but reappeared in the F_2 generation. The offspring in the F_2 generation had four phenotypes in the following proportion: Most had round yellow seeds, some had round green seeds or wrinkled yellow seeds, and a very few had wrinkled green seeds.

Problem. What can explain the observation?

Hypothesis. Mendel's hypothesis was the basis for his second law, the law of independent assortment. Alleles assort independently when gametes are formed. The allele for round seeds (A) is dominant to the allele for wrinkled seeds (a), and the allele for yellow seeds (B) is dominant to the allele for green seeds (b). Alleles for texture and color are on different chromosomes. Since the alleles assort themselves independently when gametes are formed, crossing the F_1 generation with itself will yield offspring in the F_2 generation that have the following ratio:

9   round yellow seeds
3   round green seeds
3   wrinkled yellow seeds
1   wrinkled green seeds

Exercise 9

Table 5.   Dihybrid cross

A is the allele for round seeds and a is the allele for wrinkled seeds. B is the allele for yellow seeds and b is the allele for green seeds. The F_1 generation plants produce gametes with four possible allele combinations: AB, Ab, aB, ab.

	AB	*Ab*	*aB*	*ab*
AB	*AABB*	*AABb*	*AaBB*	*AaBb*
Ab	*AABb*	*AAbb*	*AaBb*	*Aabb*
aB	*AaBB*	*AaBb*	*aaBB*	*aaBb*
ab	*AaBb*	*Aabb*	*aaBb*	*aabb*

Prediction. If an F_1 pea plant that has dominant round and yellow seeds is self-crossed, the F_2 generation will have a ratio of 9:3:3:1. Nine round yellow seeds; three round green seeds; three wrinkled yellow seeds; and one wrinkled green seeds.

Experiment.

P generation: Cross round yellow-seeded plants with wrinkled green-seeded plants.

F_1 generation: All plants have round yellow seeds. Cross the F_1 generation with itself.

F_2 generation: 315 have round yellow seeds

108 have round green seeds

101 have wrinkled yellow seeds

32 have wrinkled green seeds

Conclusion. The next step is to submit Mendel's results to a chi-square analysis. See Table 6. The total number of seeds are obtained by adding the number of seeds of each type: $315 + 108 + 101 + 32 = 556$. Then the expected results are calculated. Of the sixteen possible offspring, 9/16 of 556 or 313 should be round and yellow, 3/16 of 556 or 104 should be round and green, 3/16 of 556 or 104 should be wrinkled and yellow, and 1/16 of 556 or 35 should be wrinkled and green. Finish calculating the chi-square value to determine if Mendel's hypothesis is supported by the results. Can you accept Mendel's hypothesis?

Table 6. Mendel's F_2 generation

Phenotype	O	E	(O-E)	$(O-E)^2$	$\dfrac{(O-E)^2}{E}$
Round, yellow	315	313			
Round, green	108	104			
Wrinkled, yellow	101	104			
Wrinkled, green	32	35			
$X^2 =$					
$p =$					

3: THE TESTCROSS

A test cross determines whether an organism that exhibits a dominant trait has one or two alleles for the trait. The organism is crossed with one that has two alleles for the recessive trait. Since alleles segregate when gametes are formed, if an unknown organism has two dominant alleles and is test crossed with one that has two recessive alleles, then all offspring will exhibit the dominant trait. If the organism has one dominant and one recessive allele and is test crossed with one that has two recessive alleles, then the offspring will be in a 1:1 ratio of dominant to recessive for that trait. It does not matter how many traits you consider. If an organism has one dominant and one recessive allele for a trait, each trait will be in a 1:1 ratio of offspring for the two traits. Therefore, a monohybrid test cross will give a 1:1 ratio of offspring for the two traits. See Table 7. A dihybrid test cross will give a 1:1:1:1 ratio of offspring for the four traits. See Table 8.

Table 7.   Monohybrid testcross
The F_1 generation crossed with a white-flowered plant. The F_1 generation produces *A* and *a* alleles in a 1 : 1 ratio. The white-flowered plant produces only an *a* allele.

	A	*a*
a	*Aa*	*aa*

Table 8.   Dihybrid testcross
The F_1 generation crossed with a white-flowered, green-seeded plant. The F_1 generation produces *AB*, *Ab*, *aB* and *ab* alleles in a 1:1:1:1 ratio. The white-flowered, green-seeded plant produces only an ab allele.

	AB	*Ab*	*aB*	*ab*
ab	*AaBb*	*Aabb*	*aaBb*	*aabb*

4: MONOHYBRID CROSS OF INDIAN CORN

PROCEDURE

1. Select an F_2 ear of corn from the side bench that is the result of an F_1 hybrid (*Aa*) self-cross (*Aa* x *Aa*) with alleles on separate chromosomes.

2. Collect data by counting and recording the kernel phenotypes in Table 9.

Table 9

Phenotype	O	E	(O-E)	(O-E)^2	$\dfrac{(O-E)^2}{E}$
Purple					
Yellow					
$X^2 =$					
p =					

3. Perform the chi-square test to compare observed results with expected results. Use Table 4 to determine the p value.

4. If the p value for the calculated chi-square is greater than 0.05, then the observed ratio is statistically close enough to the expected 3:1 phenotypic ratio.

5: DIHYBRID CROSS OF INDIAN CORN

PROCEDURE

1. Select an F_2 ear of corn from the side bench. The corn is the result of an F_1 hybrid (*AaBb*) self-cross (*AaBb* x *AaBb*) with alleles on separate chromosomes.

2. Collect data by counting and recording the kernel phenotypes in Table 10.

Table 10

Phenotype	O	E	(O-E)	$(O-E)^2$	$\dfrac{(O-E)^2}{E}$
Purple, smooth					
Purple, wrinkled					
Yellow, smooth					
Yellow, wrinkled					
$X^2 =$					
$p =$					

3. Perform the chi-square test to compare observed results with expected results. Use Table 4 to determine the p value.

4. If the p value for the calculated chi-square is greater than 0.05, then the observed ratio is statistically close enough to the expected 9:3:3:1 phenotypic ratio.

6: DETERMINATION OF THE GENOTYPE OF AN UNKNOWN

You will be given an ear of Indian corn of unknown heredity. Each ear represents one of four possible crosses:

- F_2 monohybrid cross. The result of a F_1 hybrid (*Aa*) self-cross (*Aa* x *Aa*).

- F_2 dihybrid cross. The result of an F_1 hybrid (*AaBb*) self-cross (*AaBb* x *AaBb*) with alleles on separate chromosomes.

- F_1 monohybrid test cross. The result of testcrossing an F_1 hybrid (*Aa*) with the recessive parent (*Aa* x *aa*).

- F_1 dihybrid test cross. The result of testcrossing an F_1 hybrid (*AaBb*) with the recessive parent (*AaBb* x *aabb*) with alleles on separate chromosomes.

You will make a hypothesis about the genotypes of the plants that produced the ear of corn. You will collect data and submit the data to a chi-square analysis to see if the data support your hypothesis. Every kernel on the Indian corn is a seed, the result of a fusion of a male gamete (pollen nucleus) carrying one allele and a female gamete (ovum) carrying one allele.

PROCEDURE
1. **Observation**. Look for kernel color and/or texture. How many different kinds of kernels are present on your Indian corn cob? List the different kernel phenotypes and notice their proportion in the space below.

2. **Problem**. What can explain your observation? What is the genotype of the parent generation of your ear of Indian corn?

Exercise 9

3. **Hypothesis**. Your hypothesis answers the question posed by the problem. State the hypothesis you will test.

4. **Experiment**. Gather data by determining both the expected and observed numbers of kernel.

Calculate the chi-square value using the formula. Complete all calculations to three digits. Round off your answer to two digits. Determine the degree of freedom and locate the value closest to your calculated chi-square on that degrees of freedom row. Move up the column to determine the p value. See Table 11.

Table 11. Chi-square value for Indian corn

Phenotype	O	E	(O-E)	$(O-E)^2$	$\dfrac{(O-E)^2}{E}$
$X^2 =$					
$p =$					

5. **Conclusion**. The results either support or fail to support your hypothesis. State your conclusion in terms of the hypothesis. If the p value for the calculated chi-square is greater than 0.05, accept your hypothesis.

VOCABULARY

TERM	DEFINITION
Allele	
Dominant	
Recessive	
Homozygous	
Heterozygous	
Phenotype	
Genotype	

Mendel's first law	
Mendel's second law	
Monohybrid cross	
Dihybrid cross	
P generation	
F_1 generation	
F_2 generation	

108

Chi-square analysis	
Testcross	

QUESTIONS

1. A F_1 hybrid (Aa) is self crossed. What chromosome combinations will you find in the gametes of the F_1 hybrid?

2. A F_1 hybrid ($AaBb$) is self crossed. What chromosome combinations will you find in the gametes of the F_1 hybrid?

3. How can you tell whether an individual that expresses a dominant trait is homozygous or heterozygous?

4. You perform an experiment and submit your results to a chi-square analysis. Your p value is 0.08. Is your hypothesis supported?

THE MOLECULAR BASIS OF HEREDITY I
DNA STRUCTURE AND REPLICATION

OBJECTIVES

- Define the vocabulary terms from the questions at the end of this exercise.
- Recognize on a DNA model the components of a DNA molecule — the double strands, the sugar-phosphate backbones and the paired nitrogen bases.
- See on a DNA model that a purine pairs with a pyrimidine. Adenine and thymine are paired. Guanine and cytosine are paired. Note the antiparallel strands on a DNA model. One strand is from 3' to 5' and the other is from 5' to 3'.
- Observe the helical nature of the DNA molecule where each nucleotide is rotated 36° around the axis of the molecule.
- Assemble models of the four nucleotides found in a DNA molecule.
- Construct a model of a sixteen base-pair DNA molecule from component nucleotides.
- Replicate a model of a DNA molecule using the concepts of leading and lagging strands and Okazaki fragments.

TOPICS

1: EXAMINATION OF THE DNA MODEL
2: DNA SYNTHESIS
3. DNA REPLICATION

INTRODUCTION

Deoxyribonucleic acid (DNA) is a molecule that contains the hereditary information necessary for life. It has the unique ability to make a copy of itself, a feature that allows the information it contains to pass from one cell to another during cell division, thus preserving the continuity of life. The DNA molecule carries in its structure the instructions necessary for cellular activity. However, it does not program the functions of the cell directly. Information is passed from DNA to ribonucleic acid (RNA), and then to proteins. There are millions of different species of living organisms, and within each species there are individual variations. For example, oak trees differ from humans and humans differ from each other unless they are identical twins. These differences among and within species are the result of the unique proteins found in the cells of each organism. It is the code of information in the DNA that ultimately determines what proteins will be made, and thus what the organism will be.

DNA STRUCTURE

DNA is a polymer of nucleotides called a polynucleotide. The nucleotides of DNA are made of three parts: a five-carbon deoxyribose sugar, a phosphate group and a nitrogen base. There are four possible nitrogen bases: adenine, guanine, cytosine and thymine. The polynucleotide is formed by linking the phosphate of one nucleotide to the sugar of another, thus creating a repeating sugar-phosphate-sugar-phosphate backbone with the nitrogen bases bonded to each sugar. DNA is a double helix. It contains two polynucleotide strands wound around a central axis. A nitrogen base from one strand is paired with a nitrogen base of the other strand in a precise manner. Adenine pairs with thymine and cytosine pairs with guanine. The hereditary information in the DNA molecule is found in the sequence of nitrogen bases. DNA molecules are made of thousands and sometimes millions of nucleotides. A typical gene contains 1000 to 1500 nucleotide pairs.

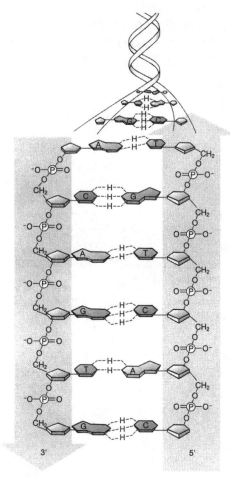

Figure 1. Two strands of a DNA double helix held together by hydrogen bonds

1: EXAMINATION OF THE DNA MODEL

The DNA model consists of two complementary polynucleotide strands, each containing sixteen nucleotides. The outside of the molecule has two sugar-phosphate backbones. Facing inside are the nitrogen bases. A nitrogen base from one strand is paired with a nitrogen base from the other strand. Work in groups of three or four.

PROCEDURE

1. **Nitrogen base pairs**. The nitrogen base pairs are found in a horizontal plane in the model. They are composed of a purine hydrogen bonded to a pyrimidine. Purines have two rings. Adenine and guanine are purines. Pyrimidines have only one ring. Cytosine and thymine are pyrimidines.

   a) Find the thymine with the brown methyl group. Thymine is bonded to adenine through two hydrogen bonds. Locate the adenine.

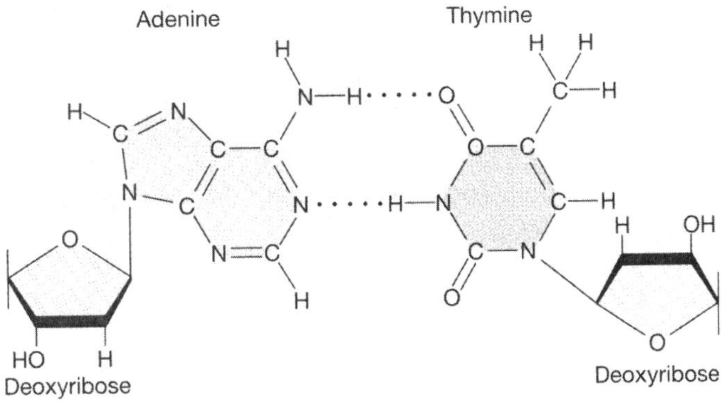

Figure 2. Adenine-thymine base pair

114

b) Find a guanine-cytosine base pair.

c) Note that there are three hydrogen bonds between the guanine and cytosine. Locate the guanine and the cytosine. Since adenine pairs with thymine and guanine pairs with cytosine, the order of nitrogen bases on one strand determines the order of nitrogen bases on the other strand. This feature is of critical importance when the DNA molecule makes an exact copy of itself.

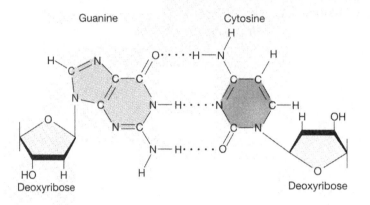

Figure 3. Guanine — cytosine base pair

2. **Deoxyribose sugar.** Locate a sugar at the top of the DNA molecule. It is composed of five carbons and one oxygen.

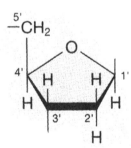

Figure 4. Deoxyribose sugar showing carbon positions

a) The #1 carbon of the sugar is linked to a nitrogen base. Count the carbons of the sugar starting with the #1 carbon.

b) Locate the top sugar on the complementary strand. Its #1 carbon is also linked to a nitrogen base.

c) Count the carbons of this sugar starting with the #1 carbon. Note that the two sugars have different carbons at the top of the molecule. One sugar has a free #3 carbon (3' carbon) and the sugar of the complementary strand has a #5 carbon (5' carbon) linked to a phosphate group.

3. **Sugar-phosphate backbone**.

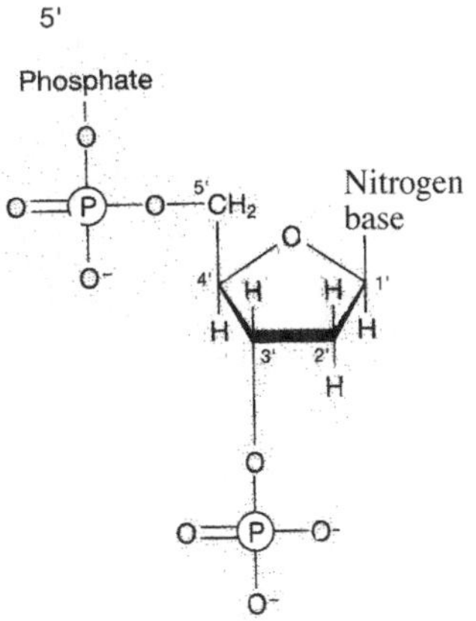

Figure 5. Sugar-phosphate backbone

   a) Locate a top sugar in the DNA molecule. Note that it is linked to a phosphate group below it.

   b) The phosphate group is then linked to a sugar immediately below it, and the second sugar, in turn, is linked to a phosphate group below it. This repeating sugar-phosphate backbone continues down the entire length of the molecule. Note that the phosphate groups link the 5' carbon of one sugar to the 3' carbon of the previous sugar.

   c) Locate the complementary top sugar in the molecule. The sugar-phosphate backbone of this strand goes to the bottom of the molecule as well. The two sugar-phosphate backbones run in opposite directions and are said to be antiparallel. One has a 5' carbon at the top and the other has a 3' carbon at the top. The sugars of the complementary strands are upside down with respect to each other.

4. Overall view of the DNA model. Each nucleotide is rotated 36° around the axis of the molecule. To check this, start with the bottom base pair and count up ten base pairs. The eleventh base pair will be facing the same direction as the first, coming full circle or 360°. Since a two-ring purine always pairs with a one-ring pyrimidine, the distance between the sugar-phosphate backbones remains the same for the length of the molecule. The hydrogen bonds between the nitrogen bases are not as strong as the covalent bonds between the other atoms in the molecule. They do give stability to the molecule, but because they are not very strong they can be broken to unzip the two strands. Look at the DNA model from the side. It has the appearance of a flexible ladder that has been twisted along its length. The sugars and phosphates form the uprights of the ladder and the nitrogen base pairs form its rungs.

2: DNA SYNTHESIS

PROCEDURE

1. Follow directions carefully! Work in groups of two.

2. Make sixteen nucleotides of each of the four nitrogen bases for a total of sixty-four nucleotides. To make a nucleotide put a phosphate group (red bead) on the 5' position of a deoxyribose sugar (white round bead). Next put one nitrogen base (adenine, guanine, cytosine or thymine) into the 1' position of the sugar.

3. Repeat until all sixty-four nucleotides are synthesized.

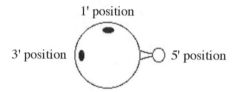

Figure 6. Carbon positions on the deoxyribose sugar (white bead)

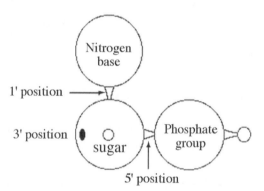

Figure 7. Nucleotide composed of a sugar, nitrogen base and phosphate group
Note 1', 3' and 5' carbon positions of the sugar

4. Sixteen nucleotides will be used to make the first strand of a DNA molecule. Select four nucleotides of each nitrogen base to make this strand.

5. Connect a phosphate group (red bead) of one nucleotide to the 3' end of a sugar (white round bead) of another nucleotide until you have a strand of sixteen nucleotides. The nucleotides can be added in any order.

Exercise 10

6. Put the strand on the laboratory table with the nitrogen bases facing up. <u>Be careful to orient the strand from the 3' end of the sugar on the left to the 5' on the right</u>. This strand will serve as a template for making its complementary strand.

7. In the space below write the order of nucleotides on the DNA strand using the letters of the nitrogen bases. Label the 3' and 5' end of the strand.

Order of nucleotides of DNA molecule
(One strand)

8. Leave the strand on the table with the 3' end on the left, the 5' end on the right and the nitrogen bases facing up. From the remaining forty-eight nucleotides bring a nucleotide with a complementary nitrogen base to the first nucleotide on the left in the original strand.

9. Orient this new nucleotide so that its nitrogen base is touching the nitrogen base of the original strand, its phosphate is to the left and its sugar 3' end is pointing to the right. Since adenine pairs with thymine and guanine pairs with cytosine, choose the complementary nucleotide. If the first nucleotide in the original strand is an adenine nucleotide, bring a thymine nucleotide. If it is cytosine, bring a guanine; if it is thymine, bring an adenine; and if it is guanine, bring a cytosine.

10. Bring a nucleotide that is complementary to the second nucleotide in the original strand, and attach its phosphate group to the 3' sugar of the first nucleotide. Continue adding complementary nucleotides until you have completed a strand that is sixteen nucleotides long. See Figure 8.

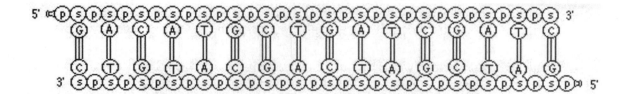

S - sugar   T - thymine
P - phosphate   C - cytosine
A - adenine   G - guanine

Figure 8. DNA segment (sixteen base pairs)

11. Note that nucleotides are added to the 3' end of the new strand. The two strands are antiparallel. The original strand goes from 3' to 5', and the new strand goes from 5' to 3'.

12. Connect the complementary nitrogen bases of the two strands with hydrogen bonds (clear connectors) to form a double-stranded DNA molecule. In the space below write the order of nucleotides in both strands. Label the 3' and 5' ends of each strand.

Order of nucleotides of DNA molecule
(Both strands)

13. Pick the molecule up and carefully twist it into a helical shape. You now have a completed DNA molecule composed of two complementary strands of sixteen nucleotides.

3: DNA REPLICATION

The DNA molecule can make an exact copy of itself, a process called *replication*. The two strands of the molecule separate and serve as templates for making new strands. Complementary nucleotides are brought to the templates, and the new strands grow by adding nucleotides to their 3' end. When replication is complete, there are two identical DNA molecules, each like the original molecule.

PROCEDURE

1. Work in groups of two. To replicate your molecule, remove the hydrogen bonds from the eight nitrogen base pairs on the right of the molecule and separate the strands (Figure 8). Since new strands grow by adding nucleotides to their 3' end, the two new strands grow in opposite directions.

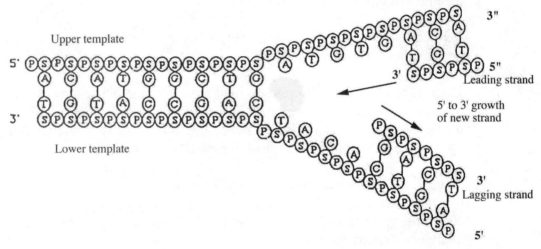

Figure 9. DNA replication

2. Using the remaining nucleotides, bring a new complementary nucleotide to the nucleotide at the 3' end of the upper template. Orient the new nucleotide so that its 3' end is pointing to the left and its nitrogen base is touching the complementary nitrogen base on the upper template. Connect the nitrogen bases of the two nucleotides with a hydrogen bond.

3. The new strand grows continuously one nucleotide at a time away from the open end in a 5' to 3' direction by adding a nucleotide to the 3' end of the sugar. The new strand is called the *leading strand*. Add nucleotides to the leading strand by attaching the 5' phosphate group of each new nucleotide to the 3' sugar of the nucleotide before it. Connect the complementary nitrogen bases with hydrogen bonds.

4. On the lower template the new strand still grows in a 5' to 3' direction, but it moves toward the open end. It is called the *lagging strand*. The lagging strand replicates in short fragments called *Okazaki fragments* that are eventually joined together.

5. Bring a complementary nucleotide to the fourth nucleotide from the 5' end of the lower strand. Orient the new nucleotide so the 3' end is to the right and connect the nitrogen bases with a hydrogen bond.

6. Replicate a short nucleotide fragment by bringing in complementary nucleotides one at a time and connecting them to the right.

7. Then go to the point where the original strands separated and repeat this process by bringing a complementary nucleotide to the eighth nucleotide from the 5' end on the lower template. Hydrogen bond three complementary nucleotides, one at a time, to the right until the fragment of new DNA is reached.

8. Separate the remaining eight nitrogen bases of the original strands by removing the hydrogen bonds. Replicate the upper strand by adding nucleotides to the 3' end of the growing strand until the left end is reached and the new strand consists of sixteen nucleotides.

9. On the lower template find the fifth nucleotide from the 3' end of the template, bring in a complementary nucleotide, one at a time, and join the nucleotides with a hydrogen bond. Move to the right and hydrogen bond three more nucleotides to the template to make another short fragment.

10. Then move to the first nucleotide on the lower template, hydrogen bond its complementary nucleotide and add the last three nitrogen bases.

11. To complete the new strand, bond the four short nucleotide fragments by joining the phosphates to the 3' end of the sugar next to them. The original DNA molecule is now replicated. There should be two identical DNA molecules that are exact copies of the original DNA.

12. Place the two DNA molecules on the laboratory table so the bottom strand of each is oriented from the 3' to 5' end from left to right. In the space below write the order of nucleotides in each DNA molecule. Label the 3' and 5' ends of each molecule.

Order of nucleotides of the original DNA molecule
(Both strands)

Order of nucleotides of the new DNA molecule
(Both strands)

13. Check to be certain that the molecules are exact copies of each other. Refer to the nucleotide sequence of the original DNA molecule that you made. These two DNA molecules should be identical to the original.

14. Disassemble the DNA molecules into their separate beads.

122

VOCABULARY

TERM	DEFINITION
DNA	
Nucleotide	
Polynucleotide	
Nitrogen bases	
Purine	
Pyrimidine	
3' and 5' ends	

Exercise 10

Antiparallel	
Okazaki fragment	
Leading strand	
Lagging strand	

QUESTIONS

1. Why is a DNA molecule called a *double helix*?

2. Where are the sugars and phosphates located on a DNA molecule?

3. Which nitrogen bases are purines? Pyrimidines?

4. What base pairs with adenine? With guanine?

5. Why are the strands of a DNA molecule antiparallel?

6. What is the rotation of the DNA helix? After how many nucleotides does the molecule have a full rotation?

7. What is the basic unit of a DNA molecule?

8. At what end of the sugar are new nucleotides added to a growing DNA molecule?

9. How is a DNA molecule replicated? What are leading and lagging strands?

EXERCISE 11

THE MOLECULAR BASIS OF HEREDITY II
PROTEIN SYNTHESIS

OBJECTIVES
- Define the vocabulary terms from the questions at the end of this exercise.
- Compare and contrast the structure of DNA and RNA.
- Describe the three types of RNA and their roles in protein synthesis.
- Learn how RNA is transcribed from a DNA template.
- Differentiate between the A and P sites on a ribosome.
- See how messenger RNA code is translated to the amino acid sequence of a protein.
- Determine the genes that code for insulin given the sequence of amino acids in an insulin molecule.
- Predict what effect mutations in the insulin genes might have on an insulin molecule.

TOPICS
1: PROTEIN SYNTHESIS
2: THE INSULIN GENES

INTRODUCTION
DNA controls the structure and function of all organisms by specifying the synthesis of proteins. The code inherent in the sequence of nitrogen bases in DNA ultimately determines the sequence of amino acids in a protein. Protein synthesis occurs in two steps. First, the nitrogen base sequence in the DNA molecule is transcribed to ribonucleic acid (RNA). And second, the nitrogen base sequence in RNA is translated into the sequence of amino acids in a protein.

Like DNA, RNA is a polynucleotide. However, RNA differs from DNA in three ways. The sugar in RNA is ribose. Ribose has one more oxygen atom than deoxyribose. There is no thymine in RNA; instead the thymine is replaced by uracil. And RNA is composed of a single polynucleotide strand. There are three types of RNA: messenger RNA (mRNA), ribosomal RNA (rRNA) and transfer RNA (tRNA). All three are transcribed from DNA.

1: PROTEIN SYNTHESIS

In this exercise you will transcribe an mRNA molecule from the DNA molecule you synthesized and then you will use that mRNA to translate a short polypeptide.

PROCEDURE: SYNTHESIS OF mRNA

1. Work in groups of two. Remake the DNA molecule you made in the previous exercise and place it on the laboratory table with the bottom strand 3' sugar on the left and the top strand 3' sugar on the right.

2. Open up the DNA molecule by removing all the hydrogen bonds (clear connectors) but keep the molecule on the laboratory table. Only the bottom 3' to 5' strand will be used as a template to transcribe the messenger RNA since RNA, like DNA, grows from a 5' to 3' direction. This DNA strand is called the *sense strand*. Remove the top strand, the antisense strand, and disassemble it.

3. Remove the last nucleotide at the 5' end of the sense strand. You should have a strand with fifteen nucleotides.

4. Add three nucleotides to the 3' end in the following order from left to right: thymine, adenine, cytosine. These three nucleotides are a start signal for the manufacture of a protein.

5. Add three nucleotides at the 5' end in the following order from left to right: adenine, thymine, cytosine. These three nucleotides are a signal to stop making the protein. The DNA strand should now be composed of twenty-one nucleotides with TAC at the 3' end and ATC at the 5' end.

6. Make six RNA nucleotides of each of four nitrogen bases for a total of twenty-four nucleotides. To make a nucleotide, put a phosphate group (red bead) on the 5' position of the ribose sugar (pink bead). Put one nitrogen base (adenine, guanine, cytosine or uracil) into the 1' position of the sugar. Repeat until all twenty-four nucleotides are synthesized.

7. To synthesize an RNA molecule coded by the DNA strand, bring complementary RNA nucleotides to the DNA template. First, bring an adenine nucleotide to the thymine at the 3' end of the DNA. Next, bring a uracil nucleotide to the adenine on the DNA. Connect the sugar of the adenine to the phosphate of the uracil.

8. Continue adding RNA nucleotides that are complementary to the DNA nucleotides until you have made an RNA molecule of twenty-one nucleotides. Note that growth of the RNA molecule occurs in a 5' to 3' direction, like the growth of DNA. This RNA strand is a messenger RNA (mRNA) that carries the genetic code of its nucleotide sequence to the ribosomes where protein synthesis takes place. The mRNA code is a triplet code. Three nucleotides (a codon) code for one specific amino acid.

PROCEDURE: SYNTHESIS OF PROTEIN

1. Work in groups of two.

2. Place the mRNA molecule on the laboratory bench with the 5' end on the left, the 3' end on the right and the nitrogen bases facing upward. Write the codons of the mRNA in the space below. Using the genetic code in Table 2, find the amino acid coded by each codon in your mRNA. Write the amino acid under each codon using the amino acid abbreviation.

Codon in mRNA   5'   AUG   _____ _____ _____ _____ _____ UAG   3'

Amino acid   f Met   _____ _____ _____ _____ _____ STOP
                              1   2   3   4   5

3. The first codon (AUG) codes for both start and the amino acid methionine. The last codon (UAG) signals stop protein synthesis. The remaining five codons in the molecule's interior code for specific amino acids. UAA, UAG and UGA all code for stop. If an interior codon specifies stop, change it to an amino acid codon.

4. Each amino acid is brought to the mRNA by a specific transfer RNA (tRNA). Transfer RNA has a loop at its base containing three central nucleotides. These three nucleotides are called the *anticodon*, and they base pair with the codon on the mRNA. A tRNA will carry only the amino acid for which it is specific.

5. Select six white twist beads and six white oval beads. The white twist beads represent amino acids. The oval beads represent tRNA.

6. Connect the beads as shown in Figure 1 to charge the tRNA with the amino acid. Charge each of the remaining five tRNA molecules (white oval beads) with an amino acid (white twist bead).

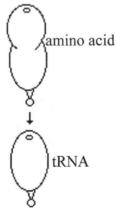

Figure 1. Charging tRNA with an amino acid

7. Put a ribosome on the laboratory table and place the mRNA strand on the ribosome with the 5' end of the mRNA on the left. Messenger RNA is read in a 5' to 3' direction. Place the first codon (AUG) in the P site and the second codon in the A site. The P site is where the polypeptide will grow. The A site is where the amino acids are brought to the ribosome.

8. Protein synthesis begins when a tRNA carrying the modified amino acid N-formylmethionyl (fMet) enters the P site. Place the peg of the fMet tRNA in the middle nucleotide (uracil) of the codon (Figure 2). In biological systems the anticodon on tRNA pairs with all three nucleotides on the mRNA codon, but for our purposes the tRNA bead can attach to the middle nucleotide of the codon only.

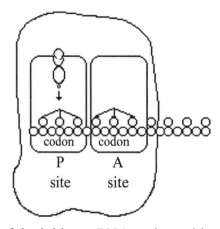

Figure 2. Bonding of the initiator tRNA-amino acid complex to the mRNA in the P site of the ribosome

9. Bring the charged tRNA that is complementary to the codon in the A site. Connect the tRNA to the middle nucleotide of the codon (Figure 3).

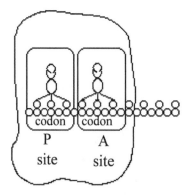

Figure 3. Bonding of the second tRNA-amino acid complex to the mRNA in the A site of the ribosome

Exercise 11

10. To form a bond between the two amino acids, remove the amino acid in the P site from its tRNA and put its peg into the amino acid in the A site (Figure 4).

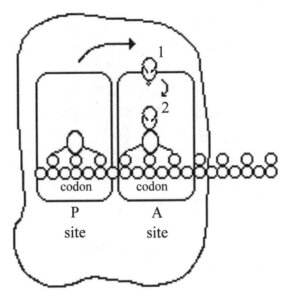

Figure 4. Transfer of the amino acid to the A site, making a short peptide of the two amino acids

11. Move the ribosome down the mRNA a distance of one codon (three nucleotides). Unsnap the uncharged tRNA previously in the P site to release it from the mRNA.

12. The second codon is now in the P site. The third codon in the A site is now ready to base pair with the anticodon of its proper tRNA. Bring this charged tRNA to the A site and connect it to the middle nucleotide of the codon (Figure 5).

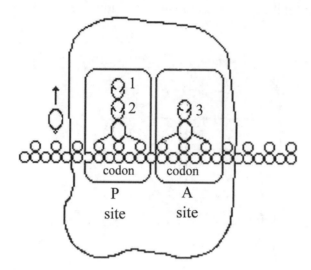

Figure 5. Movement of the ribosome the distance of one codon and release of the uncharged tRNA from the P site

13. Move the dipeptide (two amino acids) from the P site to the A site and snap it into the amino acid in the A site. Move the ribosome down another codon and continue the process until you have a polypeptide of six amino acids in the P site. The codon occupying the A site is UAG. There are no complementary tRNAs for the stop codons and protein synthesis terminates. Release the polypeptide from the tRNA. Unsnap the uncharged tRNA in the P site from the mRNA.

14. You now have a short polypeptide whose amino acid sequence was determined by the sequence of codons in the mRNA, and these mRNA codons, in turn, were directed by the sequence of nucleotides in the DNA that you made. DNA is the basic molecule of heredity. Its code ultimately determines the amino acid sequence of all proteins.

15. Disassemble all beads and put them away.

Figure 6. Overview of peptide elongation

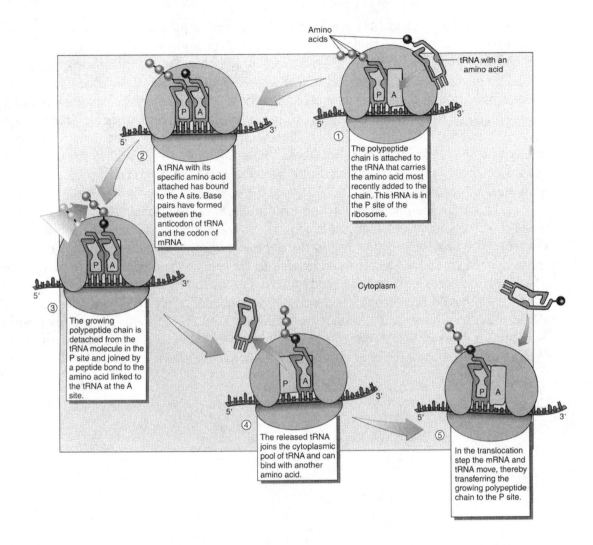

Amino acids

tRNA with an amino acid

① The polypeptide chain is attached to the tRNA that carries the amino acid most recently added to the chain. This tRNA is in the P site of the ribosome.

② A tRNA with its specific amino acid attached has bound to the A site. Base pairs have formed between the anticodon of tRNA and the codon of mRNA.

③ The growing polypeptide chain is detached from the tRNA molecule in the P site and joined by a peptide bond to the amino acid linked to the tRNA at the A site.

Cytoplasm

④ The released tRNA joins the cytoplasmic pool of tRNA and can bind with another amino acid.

⑤ In the translocation step the mRNA and tRNA move, thereby transferring the growing polypeptide chain to the P site.

2: THE INSULIN GENES

Insulin is the first protein whose entire amino acid sequence was determined. Insulin is made of fifty-one amino acids in two chains, an A chain of twenty-one amino acids and a B chain of thirty amino acids. The chains are held together by disulfide bridges. See Table 1. Insulin is a hormone that is necessary for the cellular uptake of glucose. It is used to treat insulin-dependent diabetes. The first source of insulin was the pancreas of pigs and cows. Now human insulin is produced by bacteria. Scientists are able to use the known sequence of amino acids in the insulin molecule to code backward to the nucleotide sequence in the insulin DNA. This information is used to make synthetic "genes." These "genes" are inserted into bacteria, and the bacteria make insulin in large quantities for medical use.

To determine the genes for the insulin A chain, code backward from the amino acid sequence to the DNA nucleotide sequence. The amino acid sequence of the A chain is recorded on the next page. Using the genetic code in Table 2, write the appropriate messenger RNA codons and then code back to the sense strand of the DNA. This is part of a DNA molecule that could be used to make commercial insulin.

A mutation is a change in the nucleotide sequence of a DNA molecule.
Some mutations have greater effect than others. Look at the insulin gene for chain A. Substitute a different nucleotide at nucleotide #4. Does this change the insulin molecule? What would happen to the insulin molecule if nucleotide #6 were deleted?

Sequence of amino acids in insulin chain A	Sequence of mRNA codons	Sequence of DNA codons
glycine		
isoleucine		
valine		
glutamic acid		
glutamine		
cysteine		
cysteine		
threonine		
serine		
isoleucine		
cysteine		
serine		
leucine		
tyrosine		
glutamine		
leucine		
glutamic acid		
asparagine		
tyrosine		
cysteine		
asparagine		

Exercise 11

136

Table 1.   Human insulin
            Amino acid sequence and disulfide bridges (_S__S_)

                 Chain A   Chain B

            glycine   phenylalanine
            isoleucine   valine
            valine   asparagine
            glutamic acid   glutamine
            glutamine   histidine
            cysteine   leucine
            cysteine _____S__S_____ cysteine
            threonine   glycine
            serine   serine
            isoleucine   histidine
            cysteine   leucine
            serine   valine
            leucine   glutamic acid
            tyrosine   alanine
            glutamine   leucine
            leucine   tyrosine
            glutamic acid   leucine
            asparagine
            tyrosine   valine
            cysteine_____S__S_____ cysteine
            asparagine   glycine
                                                 glutamic acid
                                                 arginine
                                                 glycine
                                                 phenylalanine
                                                 phenylalanine
                                                 tyrosine
                                                 threonine
                                                 proline
                                                 lysine
                                                 threonine

Exercise 11

Table 2.   The Genetic Code: Codons of messenger RNA
(Codons read in the 5' to 3' direction)

First position 5' end	Second position	Third position 3' end			
		U	C	A	G
U	U	UUU phenylalanine	UUC phenylalanine	UUA leucine	UUG leucine
	C	UCU serine	UCC serine	UCA serine	UCG serine
	A	UAU tyrosine	UAC tyrosine	UAA (stop)	UAG (stop)
	G	UGU cysteine	UGC cysteine	UGA (stop)	UGG tryptophan
C	U	CUU leucine	CUC leucine	CUA leucine	CUG leucine
	C	CCU proline	CCC proline	CCA proline	CCG proline
	A	CAU histidine	CAC histidine	CAA glutamine	CAG glutamine
	G	CGU arginine	CGC arginine	CGA arginine	CGG arginine
A	U	AUU isoleucine	AUC isoleucine	AUA isoleucine	AUG (start) methionine
	C	ACU threonine	ACC threonine	ACA threonine	ACG threonine
	A	AAU asparagine	AAC asparagine	AAA lysine	AAG lysine
	G	AGU serine	AGC serine	AGA arginine	AGG arginine
G	U	GUU valine	GUC valine	GUA valine	GUG valine
	C	GCU alanine	GCC alanine	GCA alanine	GCG alanine
	A	GAU aspartic acid	GAC aspartic acid	GAA glutamic acid	GAG glutamic acid
	G	GGU glycine	GGC glycine	GGA glycine	GGG glycine

Exercise 11

138

VOCABULARY

TERM	DEFINITION
RNA	
Sense strand	
Messenger RNA	
Ribosomal RNA	
Transfer RNA	
Codon	

Exercise 11

Ribosome	
A site and P site	
Mutation	

QUESTIONS

1. How do DNA and RNA differ in structure?

2. How is RNA transcribed from a DNA template?

3. How is the code in messenger RNA translated to a protein's amino acid sequence?

4. What effect would the mutations mentioned in the last paragraph of this activity have on the insulin molecule?

5. Given the sequence of amino acids in an insulin molecule, what is the sequence of nitrogen bases in DNA that codes for insulin?

Exercise 11

POPULATION BIOLOGY: THE HARDY-WEINBERG EQUILIBRIUM

OBJECTIVES
- Define the vocabulary terms from the questions at the end of this exercise.
- Explain how the Hardy-Weinberg equilibrium serves as a baseline to determine whether evolution has occurred in a population.
- List the conditions necessary to preserve the Hardy-Weinberg equilibrium.
- Given allele frequencies of a parent population, determine allele and genotypic frequencies of the next generation when conditions are stable.
- Observe the effect of genetic drift on allele frequency.
- Observe the effect of natural selection on allele frequency.
- Calculate allele frequencies of certain traits among fellow students.
- Perform a Chi-square analysis.

TOPICS
1: INVESTIGATION OF THE HARDY-WEINBERG EQUILIBRIUM
2: GENETIC DRIFT
3: NATURAL SELECTION
4: DETERMINATION OF ALLELE FREQUENCIES IN STUDENTS

INTRODUCTION
Evolution is the change in allele frequencies of a gene from one generation to the next. Any gene can be studied as the frequency or percent of the alleles of that gene in a population. A population is a group of individuals of one species that occupy a given area at the same time, can interbreed and share a common gene pool. An allele is one of two or more alternative forms of the same gene. Alleles occupy the same site on homologous chromosomes. For example, the gene determining whether someone is albino exists in two allelic forms: the dominant pigmented allele (A) or the recessive albino allele (a). It is known that the frequency of the pigmented allele A in humans is 0.99 or 99%; the frequency of the albino allele a is 0.01 or 1%. By convention, the symbol p is the frequency of the dominant allele and q is the frequency of the recessive allele. We can say that in this case p = 0.99 and q = 0.01. Note that the sum of p + q = 1 (0.99 + 0.01 = 1.00). The sum of all frequencies must be 1.0 or 100%. If you know the value of either p or q, you can find the other value by using the equation p = 1 - q or q = 1 - p.

In diploid organisms the alleles occur in pairs, one on each homologous chromosome. Since a gamete has only one of the homologous chromosomes, it can carry only one of the alleles. Thus, when you consider any population, the percent of gametes carrying a certain allele will be the same as the frequency of the allele. As an example, 99% of human eggs and sperm will have allele A for pigmentation and 1% will have allele a for albino.

142

Assuming that gametes combine at random, the frequency of male gametes carrying allele A combining with female gametes with allele A is p x p = p^2. The frequency of male gametes carrying allele a combining with female gametes with allele a will be q x q = q^2. The frequency of a male gamete with allele A combining with a female gamete with allele a is p x q (pq), and the frequency of a male gamete carrying allele a combining with a female gamete with allele A is also p x q (pq). The total probability of Aa is 2 pq.

	p	q
p	p^2	pq
q	pq	q^2

p is the frequency of the dominant allele A
q is the frequency of the recessive allele a
p^2 is the frequency of the homozygous dominant genotype AA
2pq is the frequency of the heterozygous genotype Aa
q^2 is the frequency of the homozygous recessive genotype aa

Another way of saying this is: p^2 is the percent of individuals in a population who have two alleles for the dominant trait, q^2 is the percent of individuals in a population who have two alleles for the recessive trait and 2pq is the percent of "carriers," individuals who have one dominant and one recessive allele.

Since you know the frequency of the gametes bearing the allele for albinism (p = 0.99 and q = 0.01), you can calculate the genotypic frequency of the offspring. See Table 1.

Table 1. Frequency of genotypes of offspring when the frequency of gametes bearing an allele for albinism is known

	A p = 0.99	a q = 0.01
A p = 0.99	AA p^2 = 0.9801	Aa pq = 0.0099
a q = 0.01	Aa pq = 0.0099	aa q^2 = 0.0001

The frequency of offspring who are homozygous dominant is 0.9801. The frequency of those who are homozygous recessive and albino is 0.0001 or one in 10,000. And the frequency of heterozygotes or carriers is almost two in 100 (0.0099 + 0.0099 = 0.0198). A 2% incidence is fairly common, even though albinism is quite rare. Like the sum of the allele frequency, the sum of the genotype frequency is 1.0 (0.9801 + 0.0099 + 0.0099 + 0.0001).

Evolution is the change in allele frequencies of a gene from one generation to the next. Will the allele frequency of albinism change in later generations? It might seem that the dominant allele would become more common. Godfrey Hardy and Wilhelm Weinberg working independently proposed the equation

$$p^2 + 2pq + q^2 = 1$$

The Hardy-Weinberg equilibrium states that allele frequencies will not change and no evolution will occur from one generation to the next if certain assumptions are met:
- The population is large.
- Mutations do not occur.
- The population is isolated from other populations so there is no migration into or out of the population.
- Mating is completely random.
- No selection takes place. All genotypes have equal reproductive success.

The Hardy-Weinberg equilibrium is an important standard to monitor changes of allele frequency in a population. If allele frequency does not change from one generation to the next, then evolution does not take place. However, if allele frequencies change from one generation to the next, evolution has occurred. Using the Hardy-Weinberg equilibrium allows us to see how evolution occurs at the population level.

1: INVESTIGATION OF THE HARDY-WEINBERG EQUILIBRIUM

You will test the Hardy-Weinberg equilibrium by comparing parent and offspring allele frequencies when there is a large population with no mutation or migration, and where mating is random. You will use a plastic bead to represent an allele.

PROCEDURE
1. Use 100 beads as your gene pool with eighty beads of one color as the dominant allele and twenty beads of another color as the recessive allele. The allele frequencies for the beads are 0.8 (80/100) and 0.2 (20/100) respectively. Put the beads in the container.

2. Record the color of the beads and the allele frequency. Let *A* be the allele for the dominant trait and *a* be the allele for the recessive trait.

Allele	Bead color	Allele number (actual bead count)	Allele frequency (calculate)
A			
a			

3. State the number of gametes in the population. _____

4. State the number of alleles in the population. _____

5. State the number of diploid individuals in the population. _____

6. State the phenotype (color) and genotype (*AA*, *Aa*, or *aa*) of the following individuals:

	Phenotype (color)	Genotype (*AA*, *Aa*, *aa*)
Homozygous dominant		
Heterozygous		
Homozygous recessive		

Hypothesis.
The Hardy-Weinberg equilibrium is your hypothesis: Both allele frequencies and genotypic frequencies will remain in equilibrium from one generation to the next in a large population where there is no mutation or migration, mating is random and no selection takes place.

Prediction.
Using the Hardy-Weinberg equilibrium, predict the allele frequencies and genotypic frequencies in future generations.

Experiment.

<u>PROCEDURE</u>
1. Before you begin the experiment, note the expected genotypic frequencies of the new generation genotypes *AA*, *Aa*, and *aa* when the frequency of allele *A* = 0.8 and allele *a* = 0.2 in the parent generation.

	A p = 0.8	*a* q = 0.2
A p = 0.8	*AA* $p^2 = 0.64$	*Aa* pq = 0.16
a q = 0.2	*Aa* pq = 0.16	*aa* $q^2 = 0.04$

2. Calculate the frequency of the *A* and *a* alleles in the new generation. Assume that a random assortment of gametes in the parent generation produces fifty offspring in the new generation. There will be a total of 100 alleles in the fifty offspring since each offspring has two alleles.

-   Total number of *A* alleles in the new generation
    64% of the 50 offspring = 32 are *AA*.
    Since each *AA* offspring has two *A* alleles, the *AA* offspring have a total of sixty-four *A* alleles.

    32% of the 50 offspring = 16 are *Aa*.
    Since each *Aa* offspring has one *A* allele, the *Aa* offspring have a total of sixteen *A* alleles.

-   Total number of *a* alleles in the new generation
    32% of the 50 offspring =16 are *Aa*.
    Since each *Aa* offspring has one *a* allele, the *Aa* offspring have a total of sixteen *a* alleles.

    4% of the 50 offspring = 2 are *aa*
    Since each *aa* offspring has two alleles, the *aa* offspring have a total of four *a* alleles.

-   Allele numbers in the new generation
    A = 80 (64 *A* from *AA* + 16 *A* from *Aa*)
    a = 20 (16 *a* from *Aa* + 4 *a* from *aa*)

Exercise 12

- Allele frequencies in the new generation
    $A =$ $80/100 = 0.80$
    $a =$ $20/100 = 0.20$

Note that the allele frequencies of the new generation are the same as the allele frequencies of the parent generation.

3. Now you are ready to start your experiment. Your parent population will produce fifty offspring. Remove two beads from the 100 beads without looking. The two beads are a diploid offspring of the next generation. Record the genotype (*AA*, *Aa* or *aa*) by placing a mark in Table 1.

Table 1. Number of times genotype appears in offspring

AA	
Aa	
aa	

4. Put the beads back in the gene pool so that the allele frequency and the size of the parent population remain constant. Mix the beads well.

5. Repeat steps 3 and 4 forty-nine times so you have fifty individuals in the new generation. Place a mark by *AA*, *Aa* or *aa* each time you draw that genotype from the gene pool.

6. Determine the total number of *A* and *a* alleles of the new generation and record in Table 2.

- Total number of *A* alleles in the new generation
    Number of *AA* individuals ____
    Since each *AA* offspring has two *A* alleles, the *AA* offspring have a total of ____*A* alleles.

    Number of *Aa* individuals ____
    Since each *Aa* offspring has one *A* allele, the *Aa* offspring have a total of ___ *A* alleles.

- Total number of *a* alleles in the new generation
     Number of *Aa* individuals _____
     Since each *Aa* offspring has one *a* allele, the *Aa* offspring have a total of ____ *a* alleles.

     Number of *aa* individuals _____
     Since each *aa* offspring has two *a* alleles, the *aa* offspring have a total of ____ *a* alleles.

- Allele numbers in the new generation
     $A =$___ (___ *A* from *AA* + ___ *A* from *Aa*)
     $a =$ ___ (___ *a* from *Aa* + ___ *a* from *aa*)

- Allele frequencies in the new generation
     A = ___/100 = ___
     a = ___/100 = ___

Table 2. Observed allele numbers

Number of *A* alleles	
Number of *a* alleles	

7. The Hardy-Weinberg equilibrium states that allele frequency and genotypic frequency will remain the same from one generation to the next. Therefore, of the 100 total alleles in the new generation you would expect eighty to be allele *A* and twenty to be allele *a*. Your observed allele *A* and *a* numbers in the new generation are probably not exactly like what you would have expected. To determine if the difference between the observed and expected allele numbers is due to chance or some other cause, use a chi-square analysis of the observed allele numbers. Chi-square is a statistical test that is used to determine whether the observed data are close enough to the data expected by a hypothesis to be acceptable. Chi-square tells you how many times out of 100 a deviation observed from the expected results is due to chance alone. Another way to say this is chi-square is the probability (expressed in percent) that chance alone has caused the deviation from the expected results. If we can say that chance has caused the difference between observed and expected results, then we can say that our results support our hypothesis.

148

The formula for calculating chi-square (X^2) is:

$$X^2 = \sum \frac{(O - E)^2}{E}$$

Chi-square is the sum of the squares of the difference between observed (O) values and expected (E) values, $(O - E)^2$, divided by the expected values in every category.

8. Calculate chi-square in Table 3.

Table 3. Chi-square value of observed and expected allele numbers

Allele numbers from new generation	O	E	(O-E)	$(O-E)^2$	$\frac{(O-E)^2}{E}$
A allele		80			
a allele		20			
$X^2 =$					

To interpret the chi-square value, you first must establish the degrees of freedom. Degrees of freedom are equal to the number of categories minus one. In this example, there are two categories, A and a, so there is one degree of freedom.

Look at the chi-square table (Table 4). Down the left hand side of the table are listed the degrees of freedom, and across the top of the table are probability values. In this particular application of the chi-square statistical analysis, the probability value that will be used to establish a critical chi-square value is 0.05. In statistical jargon, a probability value of 0.05 means that you are 95% sure that the effect that you are analyzing is not due to random chance.

Travel down the table on the left hand degrees of freedom column to the row with one degree of freedom. Travel across the table on the probability row to the column with a probability value of 0.05. The corresponding value is 3.84. This establishes a critical chi-square value.

When you compare your calculated chi-square value to the critical chi-square value from the table, one of two things will occur:

1) If your calculated chi-square value is smaller than the critical value, the difference between your observed and expected results is due to random chance. Your hypothesis is supported by the results.

2) If your calculated chi-square value is equal to or larger than the critical chi-square value, there is a significant difference between your observed and expected results. The difference between your observed and expected results is due to a factor other than random chance. Your hypothesis is not supported by the results.

Table 4. Chi-square table

Degrees of freedom	PROBABILITY									
	0.95	0.90	0.80	0.70	0.50	0.30	0.20	0.10	0.05	0.01
1	0.004	0.02	0.06	0.15	0.46	1.07	1.64	2.71	3.84	6.64
2	1.10	0.21	0.45	0.71	1.39	2.41	3.22	4.60	5.99	9.21
3	0.35	0.58	1.01	1.42	2.37	3.66	4.64	6.25	7.82	11.34
4	0.71	1.06	1.65	2.20	3.36	4.88	5.99	7.78	9.49	13.28
Hypothesis	Accept								Reject	

Conclusion.

Is your hypothesis supported by the results? Are the observed allele numbers of the new generation similar enough to the expected allele numbers so you can say that the Hardy-Weinberg equilibrium is supported? Has evolution taken place between the old and new generations? If the probability value for the calculated chi-square is greater than 0.05, accept your hypothesis. The deviation is small enough that chance alone accounts for it. If the probability value is equal to or less than 0.05, reject your hypothesis, and conclude that some factor other than chance is causing the deviation to be so great. For example, a probability value of 0.01 means that there is only a 1% chance that this deviation is due to chance alone.

2: GENETIC DRIFT

Genetic drift is a random change in the frequency of alleles simply by chance. Disasters can cause huge reductions in a population. When a gene pool is markedly reduced, the frequency of the alleles that survive can be very different from the frequency of the alleles of the original population. Genetic drift can cause allele frequencies to change in one generation. Genetic drift occurs when a population goes through a bottleneck. Only a few alleles survive. Depending on which alleles pass through the bottleneck, they can either be reduced or increased in frequency or sometimes even be lost or fixed in the population. An allele is said to be fixed if it is the only allele for a particular gene in the population. You will examine the effects of a bottleneck in this exercise.

Hypothesis.
Make a hypothesis stating what will happen to allele frequencies of later generations when a population goes through a bottleneck.

Prediction.
State a prediction based on your hypothesis.

Experiment.

PROCEDURE
1. Use 100 beads as your gene pool with sixty beads of one color as the dominant allele and forty beads of another color as the recessive allele. The allele frequencies for the beads are 0.6 and 0.4 respectively. This is the 0 generation.

2. Without looking, remove two beads. Repeat four times so you have ten beads representing five diploid individuals in generation 1.

3. Determine the allele frequency of generation 1 and record in Table 5.

4. In order for generation 1 to breed, you have to restore the total number of beads to 100. Calculate the number of beads of each color you will need based on the allele frequency of generation 1. For example, if you pick three beads of one color and seven beads of another color in generation one, you will need thirty beads of the first color and seventy beads of the other color to have a total of 100 beads for generation 1.

5. Again, without looking, remove two beads. Repeat four times so you have ten beads in the second generation. Determine the allele frequency of generation 2 and record in Table 5. Restore the gene pool to 100 using the new allele frequencies of this generation in order to remove ten beads for the third generation.

6. Continue this process for a total of twenty-five times or until one of the alleles is lost from the gene pool and the other is fixed. How many generations does it take before one allele is lost and the other is fixed?

7. Compare your results with other teams. Note that an allele can increase or decrease in number since the change is purely by chance.

8. As the allele frequency changes, what will happen to the genotypic frequency?

Conclusion.

PROCEDURE
1. State your conclusion based on your hypothesis.

2. Has evolution occurred?

Table 5. Changes in allele frequencies when a population goes through a bottleneck

Allele frequencies		
Generation	p	q
0	0.60	0.40
1		
2		
3		
4		
5		
6		
7		
8		
9		
10		
11		
12		
13		
14		
15		
16		
17		
18		
19		
20		
21		
22		
23		
24		
25		

Exercise 12

3: NATURAL SELECTION

Natural selection is different from genetic drift. In genetic drift the alleles survive or are lost simply by chance. In natural selection, one allele of a given gene has an advantage. Individuals with this allele are more able to survive and reproduce. Thus, the number of offspring with this allele will increase, and the offspring without this allele will decrease in future generations. It is the environment that selects which allele has the advantage. Assume that homozygous recessive individuals (*aa*) have a lethal trait and die at a young age. Individuals who are homozygous dominant (*AA*) or heterozygous (*Aa*) do not exhibit the trait and appear normal. You will remove allele *a* every time you find an individual who is homozygous recessive (*aa*).

Hypothesis.
Make a hypothesis stating what will happen to allele frequencies of later generations when there is a selective pressure against the homozygous recessive condition.

Prediction.
State a prediction based on your hypothesis.

Experiment.

PROCEDURE

1. Use 100 beads as your gene pool with sixty beads of one color as the dominant allele and forty beads of another color as the recessive allele. The allele frequencies are 0.6 and 0.4 respectively. This is the 0 generation.

2. Without looking, remove two beads.

3. Repeat four times so you have five diploid individuals in the first generation.

4. Because homozygous recessive offspring won't survive to reproduce, discard the recessive beads any time an individual has two recessive beads (homozygous recessive). Do not restore the gene pool to 100 beads. The recessive beads that you discarded are permanently lost.

5. After you remove the recessive beads, calculate the allele frequencies in the remaining population. For example, if one homozygous recessive individual occurs in generation 1, those two alleles are removed from the gene pool. The population now consists of ninety-eight beads, not 100 beads. Thus, in generation 1, the frequency of the recessive allele (q) is $38/98 = 0.387$. The frequency of the dominant allele (p) is $1 - 0.387 = 0.613$. Record the calculated allele frequencies for generation 1 in Table 6.

6. Again, without looking, remove two beads. Repeat four times so you have five diploid individuals in the second generation. Discard the recessive beads any time an individual has two recessive beads. After you remove the recessive beads, determine the allele frequency of the remaining beads in generation 2 and record in Table 6. Do not restore the gene pool to 100 beads.

7. Continue for twenty-five generations or until the recessive allele is lost in the ten beads you pick.

8. As the allele frequencies change, what will happen to the genotypic frequencies?

9. Compare your results with those of other teams.

Table 6. Changes in allele frequencies when natural selection acts on a population

Allele frequencies		
Generation	p	q
0	0.60	0.40
1		
2		
3		
4		
5		
6		
7		
8		
9		
10		
11		
12		
13		
14		
15		
16		
17		
18		
19		
20		
21		
22		
23		
24		
25		

Conclusion.

PROCEDURE

1. State your conclusion based on your hypothesis.

2. Has evolution occurred?

Exercise 12

4: DETERMINATION OF ALLELE FREQUENCIES IN STUDENTS

You will calculate allele frequencies for three traits: mid-digital hair on the fingers, tongue rolling and thumb hyperextension. For our purposes, we will assume that the three traits exhibit complete dominance. However, several have a more complicated inheritance pattern.

The frequency of the recessive phenotype is the only information you need to know to be able to determine the frequency of carriers for the recessive allele. You can calculate the frequency of the recessive phenotype by dividing the number of students with that phenotype by the number of students in the class. The frequencies of the recessive phenotype and the recessive genotype are identical because only a homozygous recessive individual expresses the recessive trait. Thus, the frequency of the recessive phenotype is equal to the recessive genotype q^2. For example, if 16% of the class has the recessive phenotype for no mid-digital hair on the fingers, you can determine p and q values for the trait. The frequency of carriers for the trait is 2pq. See Table 7.

Table 7. Calculations using the Hardy-Weinberg equation

Observation	16% (0.16) of the class has the recessive phenotype
Translation of the observation to the Hardy-Weinberg equation	$q^2 = 0.16$ (homozygous genotypic frequency)
Calculate q (recessive allele frequency)	q = the square root of 0.16 = 0.40
Calculate p (dominant allele frequency)	p = 1-q = 1 - 0.40 = 0.60
Calculate 2pq (heterozygous carrier genotypic frequency)	2pq = 2(0.60)(0.40) = 0.48

PROCEDURE

1. Poll the class for the three traits. Note that d = dominant trait and r = recessive trait. Record the number and frequency in Table 8.

2. Calculate frequencies for the dominant and recessive alleles of each trait, p and q. Calculate the frequency of carriers for each trait, 1pq. Record in Table 8.

Table 8. Frequency of alleles and carriers for three traits

Trait		# in class	% of class	q^2	q	p (1-q)	pq	2pq
Mid digital hair	d=hair present							
	r=no hair							
Tongue rolling	d=rolling							
	r=nonrolling							
Thumb hyperextension	d=can't be bent backward 60°							
	r=can be bent backward 60°							

158

VOCABULARY

TERM	DEFINITION
Evolution	
Population	
Allele	
Hardy-Weinberg equilibrium	
Genotype	
Phenotype	
Chi-square analysis	
Genetic drift	
Bottleneck	
Fixed allele	
Natural selection	

Exercise 12

QUESTIONS

1. How does the Hardy-Weinberg equilibrium serve as a baseline to determine whether evolution is occurring in a population?

2. List the conditions necessary to preserve the Hardy-Weinberg equilibrium.

3. What effect can genetic drift have on the allele frequencies of future generations?

4. If there are two alleles for a given gene and their frequencies are both 0.5, can you say which of the two alleles might increase because of genetic drift?

5. Will a significant change in allele frequency have an effect on genotypic frequency?

6. What effect does natural selection have on allele frequency? On genotypic frequency?

7. The frequency of cystic fibrosis is 0.000484, or about one in 2000 people. Calculate the allele frequencies. What proportion of people are carriers (heterozygous) for cystic fibrosis? The cystic fibrosis allele is recessive.

8. 16% of a population is homozygous recessive for a trait. What are the frequencies of the dominant and recessive alleles of that trait?

162

9. The frequency of a recessive disease is 9%. What is the frequency of the allele that causes the disease?

What is the frequency of the dominant allele?

What is the frequency of the carriers (heterozygotes) of the disease?

10. What are the frequencies of allele A and allele a in a population that has individuals with the following genotypes:

$AA = 81$
$Aa = 18$
$aa = 1$

SPECIES IDENTIFICATION AND SYSTEMATICS

OBJECTIVES
- Identify tree species by using a dichotomous key.
- Recognize how systematics provides a better understanding of phylogenetic relationships of species.

TOPICS
1: SPECIES IDENTIFICATION USING A DICHOTOMOUS KEY
2: SYSTEMATICS. CONSTRUCTION OF A PHYLOGRAM

INTRODUCTION

Although biologists have been cataloging species for over 200 years, they do not know how many species currently inhabit the earth. To date biologists have discovered, described and classified 1.5 million species. Recent estimates of the actual total number of species ranges from 5 million to 400 million. Why is it important to know how many species exist on earth? Biologists now understand that environmental stability of the world's habitats is dependent upon the multiple interactions of many kinds of species. Continued human survival is dependent upon living with and using many different species extensively and wisely. Unfortunately we are losing species faster than we are discovering them. Current estimates of the loss of tropical animal and plant species before the year 2000 range as high as 40%. The value of those species to our understanding the environment will be lost forever. There is, therefore, a real need for us to inventory as much of the species diversity of this planet as we can before it is too late. Biologists who specialize in systematics, the study of the kinds and diversity of organisms and of any and all relationships among them, have recognized the threat to the world's biodiversity by establishing *Systematics Agenda 2000: Charting the Biosphere*, which is undertaking a comprehensive inventory of all species on the earth.

Because there are so many species, some strategy is required to arrange these species into categories and classify them. Most biologists agree that the evolved structural similarities and differences among organisms should form the basis for such classifications. Thus, most classifications will attempt to reflect the evolutionary or phylogenetic relationships known to exist among organisms.

In classifying organisms all six Kingdoms are divided into major groups called *phyla* or *divisions*; each phylum or division has a name, and its members have many structural characteristics in common. Classification does not stop with phyla or divisions. Each is further subdivided, on the basis of structural characteristics, into groups called *classes*; each class has a name and certain structural characteristics in common. Classes are further divided into orders, orders into families, families into genera and genera into species.

Below are the classifications, using the previously mentioned categories, for the red oak, the wolf and the human.

Taxonomic Category	Red oak	Wolf	Human
Kingdom	Plantae	Animalia	Animalia
Phylum	Tracheophyta	Chordata	Chordata
Class	Angiospermae	Mammalia	Mammalia
Order	Fagales	Carnivora	Primates
Family	Fagaceae	Canidae	Hominidae
Genus	*Quercus*	*Canis*	*Homo*
Species	*rubra*	*lupus*	*sapiens*

The classification of any group of organisms is partly a matter of judgment on the part of the systematist. For example, some systematists prefer to combine species into a small number of large groupings, and are termed "lumpers." Others prefer to have a large number of groups containing few species, and are termed "splitters."

1: SPECIES IDENTIFICATION. USING A DICHOTOMOUS KEY TO IDENTIFY TREE SPECIES

Suppose you find yourself in a new location. How would you identify all the species of plants or animals that are present? One solution would be to use a dichotomous key. It is a useful tool for identifying species of organisms unknown to you. In this laboratory you will use a dichotomous key to identify cones and fruits of trees found in Tennessee. At each step of the key you make a choice between two alternatives for a characteristic. As you continue making choices, you eliminate species until you arrive at a final identification.

PROCEDURE

1. Use the dichotomous key on the following pages to identify the cones and fruits found on the side bench in the laboratory.

2. Record each genus and species.

Cone or fruit	Common name	Scientific name
1		
2		
3		
4		
5		
6		
7		
8		
9		
10		

KEY TO TREE CONES AND FRUITS

1. Cone with overlapping scales 2
1   Fruit, not a cone. If cone-like lacking overlapping scales 6

    2. Cone scales more or less thickened (pines) 3
    2. Cone scales almost paper thin (spruces and hemlocks) 5

3. Cones at least 4" long ....White pine *Pinus strobus*
3. Cones less than 4" long 4

    4. Cones unsymmetrical (lopsided), with stout prickles
    ....Table mountain pine *Pinus pungens*
    4. Cones with slender prickles ....Virginia or scrub pine *Pinus virginiana*

5. Cones less than 1" long ....Eastern hemlock *Tsuga canadensis*
5. Cones 1" long or more ....Red spruce *Picea rubens*

    6. Fruit shaped like a bean pod 7
    6. Fruit not shaped like a bean pod 8

7. Pod about 2" to 3" in length, very straight ....Eastern redbud *Cersis canadensis*
7. Pod 8" or more in length ....Honey locust *Gleditsia triacanthos*

    8. Fruit with a thin wing 9
    8. Fruit not winged 15

9. Fruit paired (double), the two parts united at the base (maples) 10
9. Fruit single, not in pairs 13

    10. Fruit red or reddish brown ....Red maple *Acer rubrum*
    10. Fruit green or yellow 11

11. Fruit wings forming an angle greater than 90°,
    nearly straight from end to end ....Norway maple *Acer platanoides*
11. Fruit wings forming an angle less than 90° 12

    12. Fruit V-shaped ....Boxelder *Acer negundo*
    12. Fruit U-shaped ....Sugar maple *Acer saccharum*

13. Wing encircling the seed cavity ....American elm *Ulmus americana*
13. Wing terminal (at end of seed cavity) 14

    14. Seed cavity 4-angled in cross section
    ....Tulip poplar or yellow poplar *Liriodendron tulipifera*
    14. Seed cavity flat in cross section ....Green ash *Fraxinus pennsylvanica*

15. Fruit made of many small units packed tightly together
    or borne in a loose cluster 16
15. Fruit solitary 19

16. Fruit cone-shaped ....Magnolia *Magnolia grandiflora*
16. Fruit not cone-shaped 17

17. Fruits borne in a cluster, bright red when fresh or dark red
   or black when dry ....Dogwood *Cornus florida*
17. Fruit round, golf ball size or smaller 18

    18. Fruit round and hard with sharp projections.....Sweetgum *Liquidambar styraciflua*
    18. Fruit round, lacking sharp projections
   ....Sycamore or plane tree *Platanus occidentalis*

19. Fruit an acorn (oaks) 20
19. Fruit not an acorn 25

    20. Acorn about 1/2" long ....Willow oak *Quercus phellos*
    20. Acorn longer than 1/2" 21

21. Cup conspicuously fringed at its edge ....Bur oak *Quercus macrocarpa*
21. Cup not fringed 22

    22. Cup deep, almost covering nut ....Overcup oak *Quercus lyrata*
    22. Cup shallow 23

23. Nut tawny, cup usually elongate at base, cup scales long and relatively thin
   ....Blackjack oak *Quercus marilandica*
23. Nut brown, cup usually round at the base, scales warty in appearance.......................24

    24. Nut narrow and oblong, thin spike at the tip ....White oak *Quercus alba*
    24. Nut wide, thick spike at the tip ....Red oak *Quercus rubra*

25. Nut shiny dark brown with one light spot ....Buckeye *Aseculus glabra*
25. Nut not shiny dark brown and no light spot 26

    26. Husk covering nut without seams ....Black walnut *Juglans nigra*
    26. Husk covering nut splits along definite seams 27

27. Husk prickly 28
27. Husk not prickly 29

    28. Nut rounded in cross section, more than 2" in diameter, spine of husk branched
   and needle sharp ....American chestnut *Castanea dentata*
    28. Nut triangular in cross section, less than 1" long, spines of husk weak, not
   branched..American beech *Fagus grandfolia*

29. Husk ridged at seams ....Bitternut hickory *Carya cordiformis*
29. Husk not ridged, either smooth or slightly ridged along seams
   ....Pignut hickory *Carya glabra*

2: SYSTEMATICS. CONSTRUCTION OF A PHYLOGRAM

One important objective for biologists is to determine the phylogenetic relationships among species. A phylogram reveals which species descended from other species or, alternatively, which species share immediate common ancestors.

In constructing a phylogram the systematist first observes the structural differences and similarities that exist among organisms. It should be noted that such an analysis is based on a fundamental assumption: when two species are found to share similar characteristics it is assumed that these similar characteristics were acquired from a similar characteristic in a common ancestor. If this assumption is correct, then it is possible to state that there is a phylogenetic relationship between these two species. Similar characteristics in two species that are inherited from a common ancestor are spoken of as being homologous. For example, giving birth to live young in cows and dogs is considered to be a homologous characteristic because they inherited this trait from a common ancestor. A systematist determines the organisms with the largest number of homologous characteristics in common and then constructs a phylogram to organize the species into larger groupings.

Suppose you have to construct a phylogram to represent the ancestor descendent relationships of five vertebrate animals found in the surrounding area. First, you list characteristics that indicate structural similarities and differences that exist for these organisms. Next, you indicate with a "+" if the organism possesses the characteristic and a "0" if it does not possess the characteristic.

Table 1. Summary of characteristics of vertebrate animals

	Characteristics			
	Gives birth to live young	Walking legs present	Body with hair	Warm blooded
Bass	0	0	0	0
Lizard	0	+	0	0
Duck	0	+	0	+
Cow	+	+	+	+
Dog	+	+	+	+

Now compare each organism to one another and indicate the number of shared "+" characteristics between each pair.

Table 2. Total number of shared characteristics

Total number of shared characteristics			
bass, lizard=0	lizard, duck=1	duck, cow=2	cow, dog=4
bass, duck=0	lizard, cow=1	duck, dog=2	
bass, cow=0	Lizard, dog=1		
bass, dog=0			

A phylogram is constructed by joining first the most similar pair(s), which in our example turns out to be the cow and the dog because they share four "+" characteristics. A bifurcating branch connects this pair:

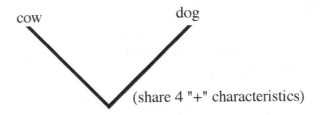

(share 4 "+" characteristics)

Note that if two pairs of organisms shared the same number of characteristics, for example, hypothetical organisms a, b sharing four "+" characteristics and hypothetical organisms d, e sharing four "+" characteristics, they would have been simultaneously joined as follows:

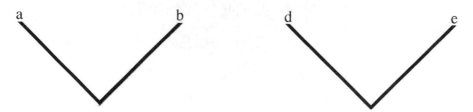

In addition, if three organisms x, y and z are equally similar among themselves, for example, if x, y share three "+" characteristics and x, z share three "+" characteristics and y, z share three "+" characteristics, then the three are joined by a trifurcating branch:

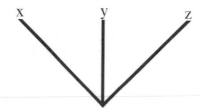

In the next step you now compare the unconnected organisms to each other and to the (cow, dog) branch. You measure the number of shared "+" characteristics between an unconnected organism to an established branch such as the (cow, dog) by finding the largest number of shared "+" characteristics between the unconnected organisms and either the cow or the dog. In this step you find that the most similar pair(s) is the duck, (cow, dog) sharing two "+" characteristics. You now connect the duck to the (cow, dog) branch by a new bifurcating branch.

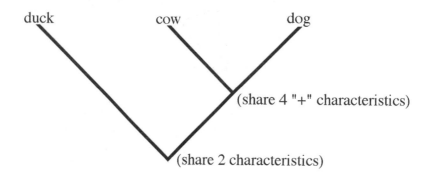

Now you repeat the above step until branches connect all organisms. Thus, in the next step you find that the most similar pair is the lizard (duck (cow, dog)) branch sharing one "+" characteristic. The lizard is then connected to the (duck (cow, dog)) branch by a new bifurcating branch.

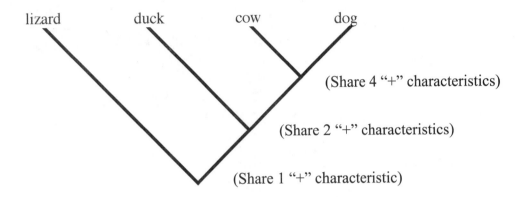

The final step connects the last remaining unconnected organism, the bass, to the (lizard (duck (cow, dog))) branch to form the completed phylogram.

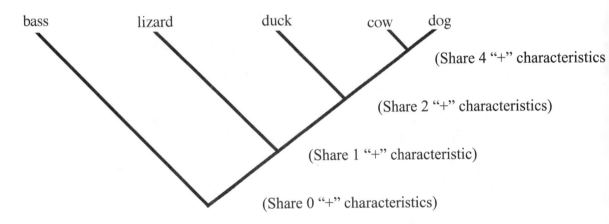

Constructing a phylogram with a large number of organisms or a large number of characters can be quite laborious. Systematists, therefore, rely on computers whenever they can to shorten the effort involved in computation. In the remaining part of this exercise you will construct a phylogram of five tree cones and fruits.

PROCEDURE
1. Examine the fruits and cones numbered 1–5 carefully.

2. Fill in the boxes below by putting a "+" showing that the characteristic is present or a "0" showing that the characteristic is absent.

Table 3. Summary of characteristics of tree fruits and cones

	Characteristics					
	Cone	Fruit	One seed	Two seeds	Nut	Nut with cap
1						
2						
3						
4						
5						

3. Compare the five tree species to one another and indicate the number of "+" characteristics between each pair.

Table 4. Total number of shared characteristics

Total number of shared characteristics			
1, 2 =	2, 3 =	3, 4 =	4, 5 =
1, 3 =	2, 4 =	3, 5 =	
1, 4 =	2, 5 =		
1, 5 =			

4. Using the procedure described above, construct a phylogram for the five tree species.

174

5. Examine the fruits and cones numbered 6–10 carefully. Fill in the box below by putting a "+" showing that the characteristic is present or a "0" showing that the characteristic is absent.

Table 5. Summary of characteristics of tree fruits and cones

	Characteristics								
6									
7									
8									
9									
10									

6. Compare these five tree species to one another and indicate the number of "+" characteristics between each pair.

Table 6. Total number of shared characteristics

Total number of shared characteristics			
6, 7 =	7, 8 =	8, 9 =	9, 10 =
6, 8 =	7, 9 =	8, 10 =	
6, 9 =	7, 10 =		
6, 10 =			

Exercise 13

7. Using the procedure described above, construct a phylogram for these five tree species.

VOCABULARY

TERM	DEFINITION
Kingdom	
Phylum	
Class	
Order	
Family	
Genus	
Species	
Dichotomous key	
Systematics	
Phylogram	
Homologous	

QUESTIONS

1. What is a fundamental assumption that must be made when you use similar characteristics to indicate phylogenetic relationships?

2. A systematist is studying the phylogenetic relationships between five organisms. The numbers of common characteristics between the organisms are:

Organisms	Number of common characteristics
1 and 2	0
1 and 3	0
1 and 4	0
1 and 5	0
2 and 3	3
2 and 4	3
2 and 5	3
3 and 4	7
3 and 5	7
4 and 5	9

Draw a phylogram from this information.

Organism 4 and 5 share a common ancestor most recently with what organism?

KINGDOMS EUBACTERIA, PROTISTA AND FUNGI

OBJECTIVES
- Define the vocabulary terms from the questions at the end of this exercise.
- Recognize the unique characteristics of bacteria, cyanobacteria, fungi and protists.
- Identify the three shapes of bacteria.
- Examine and recognize representatives of Kingdom Protista.
- Examine and recognize representatives of Kingdom Fungi.

TOPICS
1: KINGDOM EUBACTERIA. BACTERIA AND CYANOBACTERIA
2: KINGDOM PROTISTA
    A. PHYLUM CILIOPHORA. *PARAMECIUM*
    B. PHYLUM EUGLENOPHYTA. *EUGLENA*
    C. PHYLUM CHLOROPHYTA. *SPIROGYRA*
    D. PHYLUM BACILLARIOPHYTA. DIATOMS
    E. OTHER PROTISTS
3: KINGDOM FUNGI
    A. PHYLUM ZYGOMYCOTA. *RHIZOPUS*
    B. PHYLUM BASIDIOMYCOTA. MUSHROOM
4: LICHENS

INTRODUCTION

Biologists no longer think of all living things as either plants or animals. They have constructed a classification system to reflect more accurately the diversity of life forms. In this exercise you will examine three kingdoms. The first kingdom, Eubacteria, is prokaryotic. Prokaryotic cells lack a defined nucleus or other membrane-bound organelles. The remaining kingdoms we will study, Protista, Fungi, Plantae and Animalia, are eukaryotic with a defined nucleus and other membrane-bound organelles such as mitochondria.

1: KINGDOM EUBACTERIA

Kingdom Eubacteria includes bacteria and cyanobacteria, very primitive living organisms. Fossil evidence indicates that bacteria and cyanobacteria were among the earliest organisms to develop on earth. In fact, they may be similar to ancestral types from which the other kingdoms were derived. Many bacteria and cyanobacteria present today are remarkably similar to the oldest fossil types.

Members of Kingdom Eubacteria are prokaryotic. Prokaryotic cells are very simple. Most are spherical, cylindrical or spiral in shape. Different species are identified on the basis of metabolism rather than by morphologic characteristics. Most are unicellular, but cells may form colonies or filaments composed of individual cells adhering together in a mucilaginous covering or matrix. Some bacteria are motile and move with flagella. Cyanobacteria lack flagella, but some demonstrate a gliding movement produced by secretion of mucilage by the cell.

Eubacteria reproduce asexually by binary fission; the cell divides into two identical daughter cells after the genetic material has been replicated. Eubacteria are either heterotrophic or autotrophic. Heterotrophic organisms cannot manufacture their own food so they must absorb or ingest nutrients. Heterotrophic bacteria are essential for the recycling of elements between living organisms and their environment. Cyanobacteria are autotrophs that use photosynthesis to convert carbon dioxide and water to organic matter.

PROCEDURE

1. Observe the photomicrographs of bacteria on the rear bench.

2. Draw the three different shapes of bacteria in the boxes provided.

Coccus   Bacillus

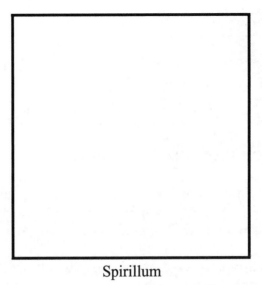

Spirillum

3. Make a wet mount of *Oscillatoria*, a filamentous cyanobacterium. To use the microscope, start with the lowest power so you will have the largest field to see the specimen. Then move to a higher power. Adjust the amount of light for the best view. Observe an area where two filaments cross. You should see a gliding movement caused by the secretion of a mucilaginous sheath. Sketch filaments of *Oscillatoria* in the space below.

Oscillatoria

182

2: KINGDOM PROTISTA

Kingdom Protista is highly diverse. It is an artificial grouping of "leftovers," not a natural, phylogenetic association. Members of this kingdom include all simple eukaryotic organisms that are neither plants, animals nor fungi. Some are autotrophic, some are heterotrophic. There is extensive variation in cell organization and life cycles.

A. PHYLUM CILIOPHORA. *Paramecium*

Members of Phylum Ciliophora are unicellular organisms that are covered with thousands of whip-like projections called cilia. Coordinated movement of the cilia enables the organism to move forward, backward, and make turns. Ciliates are heterotrophic.

Paramecium is a typical representative of Phylum Ciliophora. *Paramecium* has a specialized feeding area known as the oral groove that collects food particles such as bacteria and deposits them into food vacuoles. Asexual reproduction is by longitudinal cell division. Sexual reproduction occurs by conjugation between two cells. Two cells attach to one another and exchange genetic material. Then the cells separate leaving two cells with new genetic material.

PROCEDURE

1. Prepare a wet mount of *Paramecium*. Add a small drop of Protoslo™ before you apply the coverslip.

2. Draw the *Paramecium* in the space provided.

Paramecium

B. PHYLUM EUGLENOPHYTA. *Euglena*

Phylum Euglenophyta is composed mostly of single-celled, autotrophic organisms that move by the action of a flagellum. This group has often been used as an example of the plant-animal dilemma. The presence of chloroplasts and the ability to photosynthesize is a plant-like characteristic. Members of this group are animal-like because even though they photosynthesize, they are able to ingest food. Sexual reproduction does not occur in any member of Phylum Euglenophyta.

Euglena species are found in fresh and salt water, usually in areas with high organic content. The presence of large numbers of *Euglena* in a water sample is an indicator that organic pollution is present. Even though most are autotrophic they can also assimilate organic material heterotrophically. *Euglena* is a unicellular organism that moves by means of a single anterior flagellum that emerges from the reservoir region. A short, non-emergent flagellum is located inside the reservoir. An eyespot consisting of pigmented granules in the reservoir region is sensitive to light.

Each cell has a number of membrane-bound organelles called chloroplasts that contain the pigments involved in photosynthesis. There is a single large nucleus. The outside of the cell is bounded by a cell membrane. Since there is no rigid cell wall, cells have the ability to change shape. Reproduction occurs by division of the cell.

PROCEDURE
1. Prepare a wet mount of *Euglena*. Add one drop of Protoslo ™ before you apply the coverslip. Find the chloroplasts, the nucleus and the reddish eyespot near the anterior end. Locate the anterior, emergent flagellum and observe movement of the cell.

2. Draw the *Euglena* in the space provided and label the cellular structures.

Euglena

184

C. PHYLUM CHLOROPHYTA. *Spirogyra*

There is much variation among members of Phylum Chlorophyta, the green algae. There are many simple unicellular forms, but more complex multicellular forms exist as well. Because biochemical characteristics such as the type of pigments and storage products are identical to those in higher plants, it is generally accepted that higher plants arose from ancestral green algae.

Spirogyra is a multicellular filamentous green alga commonly found in quiet waters. The most unique characteristic is the spiral shaped chloroplast. Asexual reproduction occurs by filament fragmentation.

PROCEDURE
1. Prepare and examine a wet mount of *Spirogyra*.

2. Draw a filament in the space provided and label a chloroplast.

Spirogyra

D. PHYLUM BACILLARIOPHYTA. Diatoms

Most members of Phylum Bacillariophyta are unicellular and have cell walls composed of silica. Diatomaceous earth is composed of deposits of diatom cell walls. It is used for abrasives, filters and chalk.

Diatoms are common in both fresh water and marine habitats where they are the primary source of food for small aquatic animals. Of all of the primary producers on earth, members of this phylum fix more carbon dioxide into usable carbohydrate than any other group.

PROCEDURE
1. Examine a diatom prepared slide.

2. Draw a diatom in the space provided.

Exercise 14

Diatom prepared slide

E. OTHER PROTISTS

You have examined several representatives of Kingdom Protista. Many more examples exist and are very important ecologically and economically. On the side bench are representative specimens of some other Protista phyla.

3: KINGDOM FUNGI

Kingdom Fungi is a large and diverse group of eukaryotic organisms. All lack chlorophyll and are heterotrophs. Saprophytic forms obtain food from decomposition and absorption of nutrients from dead organic matter. Parasitic forms extract nourishment from living plants or animals. Some fungi invade plant roots to form mycorrhizae "fungus roots." The plant benefits because the fungus decomposes organic material in the soil and supplies important minerals directly to the roots of the plant. The fungus benefits from sugars and amino acids provided by the plant roots.

Some Fungi are single celled. Other Fungi form filaments called hyphae. Hyphae may be woven together to form a mycelium. The mushroom is composed of mycelia.

A. PHYLUM ZYGOMYCOTA. *Rhizopus*

Rhizopus, the common bread mold, is composed of haploid hyphae. Hyphae of this fungus grow over the surface of a food source such as bread. A prostrate hypha forms an erect sporangium that produces asexual haploid spores. The spores are released, fall on a food source, and then germinate into haploid filaments. Sexual reproduction occurs between two different mating strains. Haploid hyphae of one strain fuse with the haploid hyphae of another strain to form a diploid structure known as a zygospore. The zygospore has thick, dark walls ornamented with spines. It is resistant to adverse environmental conditions. When conditions are favorable, the zygospore undergoes meiosis and the products develop into new haploid hyphae.

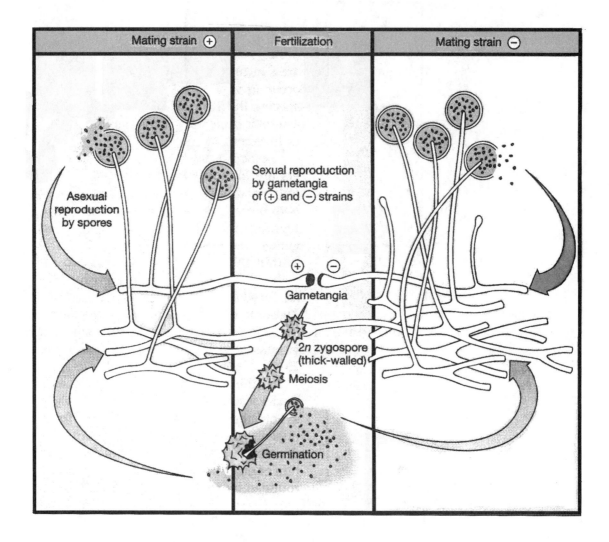

Figure 1. Life cycle of the bread mold *Rhizopus*

PROCEDURE

1. Examine the poster of *Rhizopus* reproduction on the side bench.

2. Examine a prepared slide of *Rhizopus* zygospores.

3. Using the prepared slide, make sketches of hyphae, zygospores and sporangia in the space provided.

Rhizopus prepared slide

B. PHYLUM BASIDIOMYCOTA. Mushrooms

Members of Phylum Basidiomycota produce club-shaped spore bearing structures called basidia.

Basidiomycetes are fungi composed of a mass of underground mycelia that develop into a mushroom above the ground. Mushrooms have a stem-like stipe and a cap with gills. The gills bear basidia where haploid spores are produced. Spores are released from the mushroom and germinate into haploid mycelia. Mycelia of different mating strains fuse but the nuclei do not fuse. The result is a cell with two types of haploid nuclei, a dikaryotic cell. The cell divides by mitosis to form a dikaryotic mycelium that eventually develops into a mushroom. Two haploid cells fuse in the basidium to form a zygote, the only diploid cell in the life cycle. The zygote immediately undergoes meiosis to produce haploid spores, and the cycle is repeated.

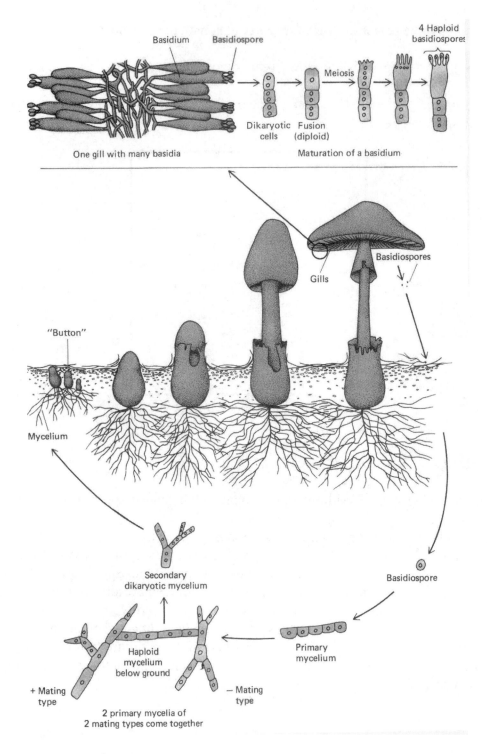

Figure 2. Life cycle of a typical mushroom

PROCEDURE

1. Examine a prepared slide of a mushroom cap and look for haploid spores.

2. Sketch what you see in the space provided.

Mushroom spores

4: LICHENS

Lichens are a symbiotic association between a fungus and either a green alga or a cyanobacterium. Algae or cyanobacteria photosynthesize and provide the fungus with nutrients. In turn, the fungus absorbs water and minerals from the environment that benefit the alga or cyanobacterium. Lichens are sensitive to air pollution; therefore they are good indicators of air quality. Look at the lichens on the side bench.

PROCEDURE

1. Observe the tree branches that have lichen growing on them. Note the different patterns of lichen growth.

2. Examine a prepared slide of a lichen. Identify the algal or cyanobacterial component and the fungal component. Draw in the space provided.

190

Lichen prepared slide

VOCABULARY

TERM	DEFINITION
Prokaryotic	
Eukaryotic	
Binary fission	
Heterotrophic	
Autotrophic	
Cilia	
Oral groove	
Flagellum	
Eyespot	
Chloroplasts	

TERM	DEFINITION
Mycorrhizae	
Hyphae	
Mycelium	
Spore	
Zygospore	
Sporangium	
Dikaryotic	
Basidium	

QUESTIONS

1. What is the distinguishing characteristic of Kingdom Eubacteria?

2. The eubacteria you saw in lab today were _____ and _____.

3. The three shapes of bacteria are _____, _____ and _____.

4. Why are autotrophic protists such as diatoms important?

5. Give an example of a heterotrophic organism studied in this exercise.

6. Give an example of an autotrophic organism studied in this exercise.

7. Give an example of an organism that can be either heterotrophic or autotrophic.

8. How do fungi obtain food?

9. How does *Rhizopus* reproduce sexually?

10. Review the life cycle of a mushroom.

11. What advantage does a mycorrhizal association have for a plant? f=For a fungus?

12. Lichens are composed of representatives of what Kingdoms?

13. Complete Table 1 on the next page by stating characteristics of the organisms.

Table 1. Characteristics of prokaryotes, protists and fungi

	Prokaryotic or Eukaryotic	Autotrophic or Heterotrophic	Motile or Nonmotile
Kingdom Eubacteria			
Oscillatoria			
Kingdom Protista			
Paramecium			
Euglena			
Spirogyra			
Diatoms			
Kingdom Fungi			
Rhizopus			
Mushrooms			

KINGDOM PLANTAE
SPORE-BEARING PLANTS
PHYLA BRYOPHYTA AND PTEROPHYTA

OBJECTIVES
- Define the vocabulary terms from the questions at the end of this exercise.
- Observe the morphology of representatives of Phylum Bryophyta and Phylum Pterophyta.
- Recognize similarities and differences between mosses and ferns.
- Explain alternation of generations.
- Diagram the life cycles of the mosses and ferns.
- Recognize the gametophytes and sporophytes of a moss and a fern.

TOPICS
1: PHYLUM BRYOPHYTA. MOSSES
2: PHYLUM PTEROPHYTA. FERNS

INTRODUCTION

Members of Kingdom Plantae are eukaryotic and multicellular. Most contain pigments that allow them to photosynthesize. Plants have a higher level of organization and specialization than protists. Spore bearing plants have an alternation of generations where they spend part of their life cycle in multicellular haploid phase and part in a multicellular diploid phase. The diploid phase is the sporophyte generation. Meiosis takes place in the sporophyte to produce haploid spores that grow into the haploid gametophyte generation. The gametophyte gives rise to haploid gametes that fuse to form the sporophyte generation, and the life cycle is repeated.

In the next two exercises you will study four phyla in the plant kingdom. You will learn distinguishing characteristics of each phylum so that you can make comparisons between them. You will use these characteristics to construct a phylogenetic tree in order to see evolutionary relationships among the plants you will study.

The phylogenetic tree will be based on three major characteristics:
- Dominant phase of the life cycle. Is the gametophyte or sporophyte dominant?
- Method of reproduction. Is the plant spore-bearing or seed-bearing?
- Type of seeds. Are the seeds naked or are they enclosed in a fruit?

198

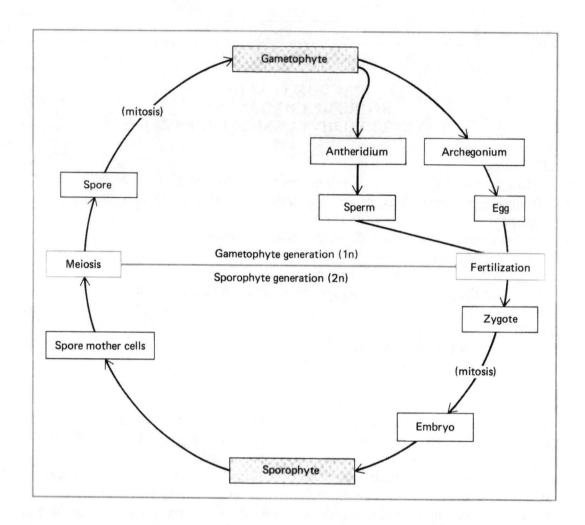

Figure 1. A general plant life cycle. Plants alternate generations, spending part of their life cycle in a multicellular diploid sporophyte generation and part of their life cycle in a multicellular haploid gametophyte generation.

1: PHYLUM BRYOPHYTA. MOSSES

Mosses are small nonvascular plants that require a moist habitat. Water is essential for sexual reproduction because flagellated sperm must swim to reach an egg. The small, leafy green plant is a haploid gametophyte. Gametophytes form multicellular structures that produce male and female gametes. The female reproductive structure is called an archegonium. The archegonium undergoes mitosis to produce non-motile eggs. The male reproductive structure is called an antheridium. An antheridium produces hundreds of small flagellated sperm that must reach the archegonium by swimming. After fertilization, the diploid zygote is retained within the archegonium and undergoes mitosis to form the sporophyte. The sporophyte is a multicellular diploid phase of the life cycle. In mosses the sporophyte grows from the archegonium and absorbs nutrients from the female gametophyte. The sporophyte is dependent on the gametophyte. Cells in the sporangium of the sporophyte undergo meiosis to produce haploid spores. Spores are released, and they germinate into a new gametophyte. The gametophyte is the dominant phase of the life cycle.

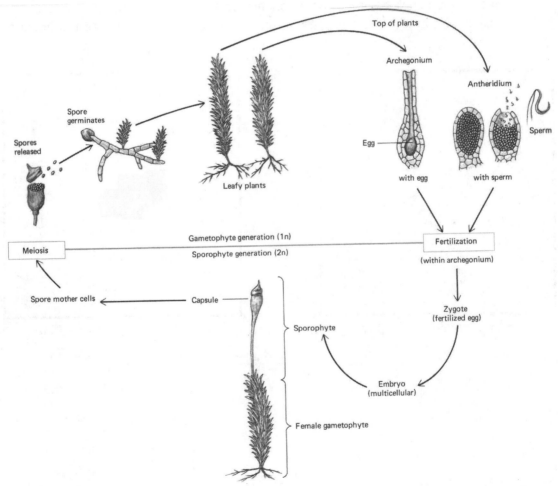

Figure 2. Moss life cycle

200

PROCEDURE

1. Observe moss reproduction in the video.

2. Examine the preserved moss specimens. Identify the male and female gametophytes and the sporophyte.

3. Draw a female gametophyte with an attached sporophyte in the space provided. Label the site where meiosis occurs to produce spores.

4. On the side bench examine the Kodachrome slide of a longitudinal section of a female moss gametophyte. See the archegonium where a non-motile egg is produced by mitosis.

5. On the side bench examine the Kodachrome slide of a longitudinal section of a male moss gametophyte. See the antheridium where motile sperm are produced by mitosis.

6. On the side bench examine the Kodachrome slide of a longitudinal section of a moss sporophyte. See the sporangium where haploid spores are produced by meiosis.

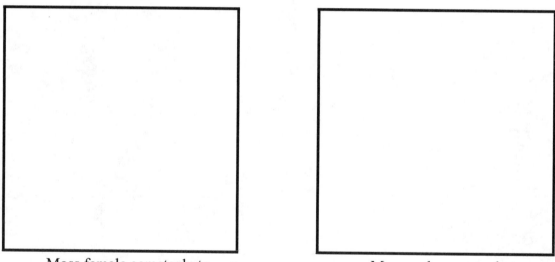

Moss female gametophyte

Moss male gametophyte

Moss sporophyte

2: PHYLUM PTEROPHYTA. FERNS

Ferns are more advanced than mosses. They exhibit true vascular tissue with both xylem and phloem extending to all parts of the plant. Xylem transports water from roots to leaves. Phloem transports the products of photosynthesis. Ferns are not as dependent on a moist habitat as mosses. However, a moist habitat is still essential for fertilization to occur because sperm are flagellated.

Ferns exhibit an alternation of multicellular generations. The sporophyte and gametophyte are independent of one another, but the dominant sporophyte is much larger and persists longer. A dominant sporophyte is an advanced evolutionary trait. The ferns common in woods and along stream banks are sporophytes. Cells in the sporangium of the sporophyte undergo meiosis to produce haploid spores. A spore germinates to form a small haploid gametophyte. The gametophytes have antheridia that produce flagellated sperm and archegonia that produce non-motile eggs. The motile sperm must swim from the antheridium to the egg in the archegonium for fertilization to occur. Following fertilization the zygote undergoes mitosis to produce a sporophyte. At first the sporophyte grows out of the archegonium of the gametophyte. As the sporophyte continues to grow, the female gametophyte withers away.

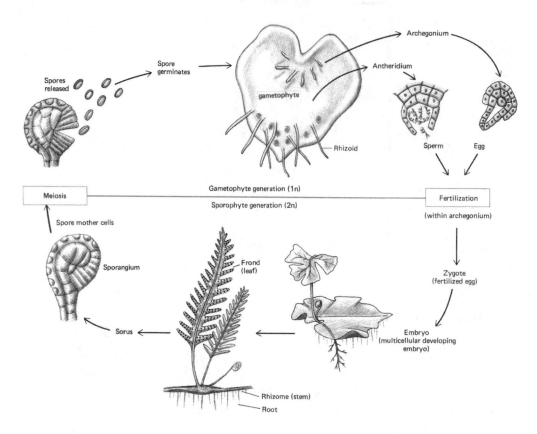

Figure 3. Fern life cycle

PROCEDURE

1. Examine the fern leaf on the side bench using a dissecting scope. This is the sporophyte generation. On the underside of the leaves you may observe small brown or black structures. Each one is a sorus containing many sporangia. Haploid spores are produced by meiosis in the sporangia. Draw the sporophyte with sori in the space provided.

Fern sporophyte

2. Examine the fern gametophytes with archegonia on the side bench. The gametophyte is small and heart-shaped. Archegonia are cup-shaped and may be found in the notch area. Draw a gametophyte with archegonia in the space provided.

3. Examine the fern gametophytes with antheridia on the side bench. Antheridia are round and are scattered on the underside of the gametophyte. Draw a gametophyte with antheridia in the space provided.

4. Examine the fern gametophytes with young sporophytes on the side bench. Draw and label in the space provided.

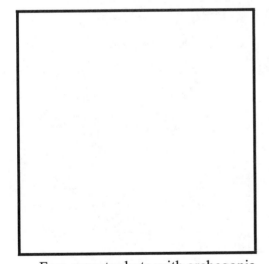

Fern gametophyte with archegonia

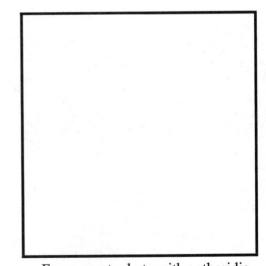

Fern gametophyte with antheridia

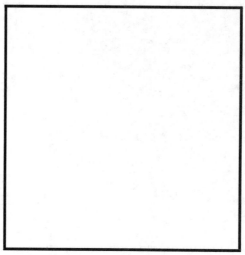

Fern gametophyte with young sporophyte

VOCABULARY

TERM	DEFINITION
Alternation of generations	
Gametophyte	
Sporophyte	
Archegonium	
Antheridium	
Spore	
Vascular tissue	
Xylem	
Phloem	
Sorus	

QUESTIONS

1. In what kind of habitat would most of these organisms be found? Explain why.

2. Review the life cycles of mosses and ferns.

3. Is the gametophyte or sporophyte dominant in mosses? In ferns?

4. Review your drawings of gametophytes and sporophytes.

5. Do the gametophyte and sporophyte generations live independently in mosses?

6. Do the gametophyte and sporophyte generations live independently in ferns?

208

7. Why is the fern considered to be more advanced than mosses?

EXERCISE 16

KINGDOM PLANTAE
PHYLUM ANTHOPHYTA
ROOTS, STEMS AND LEAVES

OBJECTIVES
- Define the vocabulary terms from the questions at the end of this exercise.
- Identify the tissues in the root.
- Identify the tissues of an herbaceous monocot stem, herbaceous dicot stem and a woody dicot stem.
- Identify the tissues in a leaf.

TOPICS
1: ROOTS
2: STEMS
   A. HERBACEOUS DICOT
   B. WOODY DICOT
   C. MONOCOT
3: LEAVES

INTRODUCTION
In this exercise you will examine roots, stems and leaves of angiosperms. Roots serve to anchor a plant to its substrate and also absorb water and minerals.

Stems support leaves and reproductive structures. Stems also provide a pathway for the transport of water up from the roots through conductive vessels called xylem. Products of photosynthesis are transported from the leaves down to the rest of the plant by the phloem.

Leaves are the main photosynthetic organs for most plants. Energy from sunlight drives chemical reactions that fix carbon dioxide from the air into carbohydrates that can be used for energy by the plant. Variations in the form of leaves permit adaptation to many different habitats.

1: ROOT

The roots of a plant absorb water and dissolved minerals. The tip of a root is covered by a protective root cap. Behind the root cap is the meristematic zone where cell division takes place. After the meristematic zone is an area where newly formed cells elongate and then begin to differentiate into specialized cells. One type of specialized cell is a root hair. Root hairs provide increased surface area to facilitate osmosis. Other specialized cells in the root include xylem and phloem.

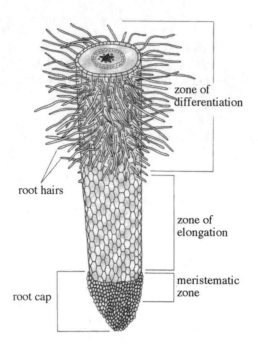

Figure 1. Root longitudinal section

<u>PROCEDURE</u>

1. Study a prepared slide of a longitudinal section of a root tip. Note the root cap and root hairs. Root hairs greatly increase the surface area of the root. Water and dissolved nutrients enter through root hairs.

2. Draw and label these areas of a root longitudinal section:
    ▪ Root cap: cells at the tip of the root that protect the apical meristem.
    ▪ Meristematic zone: area of active mitotic division found behind the root cap.
    ▪ Zone of elongation: cells formed in the meristematic zone enlarge and lengthen.
    ▪ Zone of differentiation: outer cells differentiate into root hairs for increased absorption of water.

Exercise 16

Root longitudinal section

3. Examine prepared slides of a root cross section. Find these tissues:
   - Epidermis: outermost layer of cells.
   - Endodermis: cells that surround the stele.
   - Stele: look for a central cylinder containing xylem and phloem.
     - Xylem: conducts water and minerals and provides support. The cells have thick walls and are dead at maturity.
     - Phloem: carries sugars made in photosynthesis. The cells are smaller and have thinner walls than xylem cells. At maturity phloem cells are alive.

4. Draw and label a root cross section in the space provided. Include the tissues listed above.

Root cross section

Exercise 16

2: STEMS

A. HERBACEOUS DICOT STEM

Many species of angiosperms do not become woody and instead remain herbaceous or soft. All dicots have a characteristic arrangement of vascular bundles in a ring around the perimeter of the stem. In the vascular bundle, xylem is toward the center of the stem and phloem is nearer the outside of the stem. Between the xylem and phloem is a meristematic tissue called vascular cambium. Cells in the cambium divide, then form xylem toward the inside of the stem and phloem toward the outside. Xylem cells transport water and minerals. Phloem cells transport mostly carbohydrates.

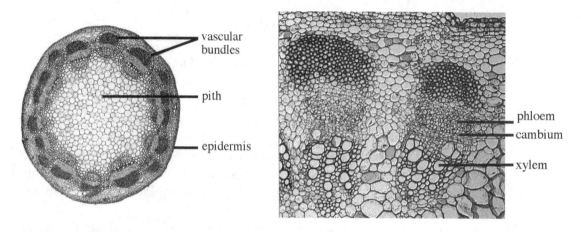

Figure 2. Herbaceous dicot stem cross section

PROCEDURE
1. Examine a prepared slide of a cross section of an herbaceous dicot. Locate the following tissues:
   - Pith: tissue occupying the central portion of a dicot stem.
   - Vascular bundles: strands of tissue containing xylem and phloem.
   - Xylem: located in a vascular bundle toward the interior of the stem.
   - Phloem: located in a vascular bundle toward the exterior of the stem.
   - Vascular cambium: located between the xylem and phloem.

2. Draw and label an herbaceous dicot stem cross section.

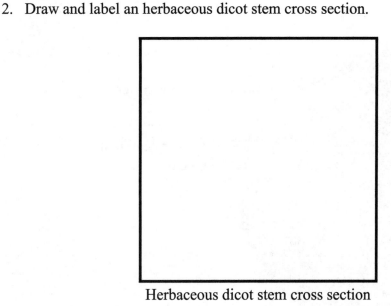

Herbaceous dicot stem cross section

B. WOODY DICOT STEM

The first year of growth in a woody dicot stem is primary growth. Growth after the first year is secondary growth. Vascular cambium, a continuous cylinder of meristematic tissue, divides to produce secondary growth. Xylem cells are formed on the inside of the cylinder of cambium, and phloem cells are formed to the outside of the cambium. Phloem from the previous years is crushed against the inside of the bark. Xylem from the previous year remains and new xylem is added. Production of new xylem causes an increase in stem diameter each year.

Pith   Primary xylem

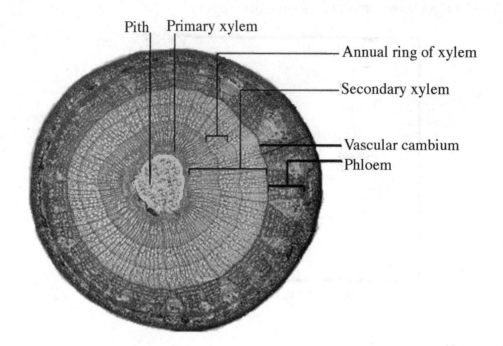

Annual ring of xylem

Secondary xylem

Vascular cambium
Phloem

Figure 3. Woody dicot stem cross section—three years

PROCEDURE

1. Examine cross section slides of a one-year-old woody dicot using a stereomicroscope.

2. Examine a cross section slide of a three-year-old woody dicot using a stereomicroscope.

3. Draw a three-year-old stem and label the secondary xylem and secondary phloem.

Woody dicot stem (three years old) cross section

2: HERBACEOUS MONOCOT STEM

In the herbaceous monocot stem, vascular bundles are scattered throughout the stem. There is no vascular cambium, thus no secondary growth.

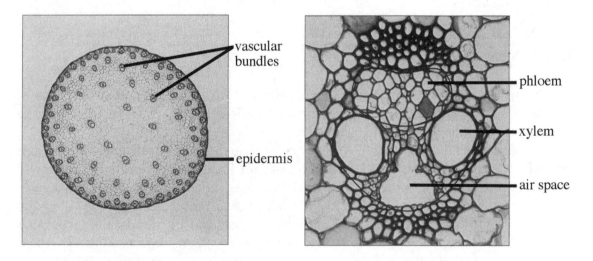

Figure 4. Monocot stem cross section

PROCEDURE

1. Study the model of a monocot stem.

2. Examine a cross section slide of a typical monocot stem.

3. Draw what you see in the space provided. Label the xylem and phloem.

Monocot stem

3: LEAVES

The primary function of leaves is photosynthesis. A secondary function of leaves is transpiration. Differences in the structure of various kinds of leaves demonstrate adaptations to a wide variety of habitats.

The outside of a leaf is covered with a waxy cuticle that helps control water loss from the leaf. The epidermis is the outermost layer of cells found on both the upper and lower surfaces of the leaf. The epidermis has small openings called stomata (singular, stoma). Guard cells control the size of the opening. When guard cells are turgid, the stoma is open. Loss of water from the guard cells by osmosis causes guard cells to become flaccid, and the stoma closes. Opening and closing of the stoma regulates the movement of carbon dioxide into the leaf and oxygen out of the leaf. The stoma also helps to regulate the loss of water by transpiration.

The mesophyll is located between the upper and lower epidermis. The palisade mesophyll is located nearest the upper epidermis. Palisade cells are stacked vertically and are photosynthetic. The tight, precise arrangement of palisade cells maximizes photosynthesis. The lower portion of the mesophyll is made of loosely arranged cells called the spongy mesophyll. These loosely arranged cells in the spongy mesophyll are well adapted to facilitate movement of gases.

The veins of a leaf run through the mesophyll. Each vein contains xylem and phloem. Xylem is closer to the upper epidermis and phloem is closer to the lower epidermis. To understand why, imagine a vascular bundle in the stem of a dicot. The xylem is toward the inside of the stem and the phloem is toward the outside. As that vascular bundle bends to go to a branch and ultimately a leaf, the inner xylem is on the top of the bending bundle and the outer phloem is on the bottom of the bundle.

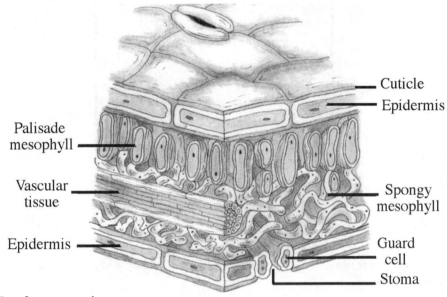

Figure 5. Leaf cross section

PROCEDURE

1. Study the model and poster of a leaf cross section.

2. Study a prepared slide of a leaf cross section. Locate the following areas:
   - Cuticle: a thin, waxy secretion from the epidermal cells, usually thickest on the upper surface of the leaf. This layer is not observable on prepared slides.
   - Upper and lower epidermis: the single layer of cells on the leaf's upper and lower surfaces.
   - Stoma: an opening in the epidermis. Stomata are most numerous on the lower surface of the leaf. The size of the opening is controlled by a pair of guard cells that shrink, thus closing the stoma, or swell, thus widening the stoma.
   - Palisade mesophyll: elongated, tightly packed cells that are the main area for photosynthesis.
   - Spongy mesophyll: an area of loosely packed, irregularly shaped cells with larger spaces between them.
   - Xylem: found nearer the upper epidermis.
   - Phloem: found nearer the lower epidermis.

218

3. Draw a leaf cross section in the space provided. Label the upper and lower epidermis, the palisade and spongy mesophylls, xylem, phloem and two guard cells.

Leaf cross section

4. Examine a prepared slide of leaf epidermis. Locate a stoma surrounded by two guard cells.

5. Draw the leaf epidermis in the space provided. Label the stoma and two guard cells.

Leaf epidermis

VOCABULARY

TERM	DEFINITION
Root hair	
Root cap	
Meristematic zone	
Zone of elongation	
Zone of differentiation	
Epidermis	
Stele	
Endodermis	
Xylem	
Phloem	

TERM	DEFINITION
Pith	
Vascular bundles	
Vascular cambium	
Primary growth	
Secondary growth	
Cuticle	
Upper and lower epidermis of the leaf	
Palisade mesophyll	
Spongy mesophyll	
Stoma	
Guard cells	

Exercise 16

QUESTIONS

1. What function do root hairs serve? Why are they important?

2. Why is the xylem normally found above the phloem in the veins of leaves? Refer to the position of xylem and phloem in the stem.

3. Review the morphology of a herbaceous dicot stem, a woody dicot stem, a monocot stem and a leaf.

KINGDOM PLANTAE
SEED PLANTS
PHYLUM CONIFEROPHYTA AND PHYLUM ANTHOPHYTA

OBJECTIVES
- Define the vocabulary terms from the questions at the end of this exercise.
- Compare the life cycles of gymnosperms and angiosperms.
- List some of the adaptations that have made angiosperms and gymnosperms successful in a terrestrial environment.
- Identify the reproductive structures of gymnosperms and angiosperms.
- Explain the reproductive processes in gymnosperms and angiosperms.
- Construct a phylogram of plant groups using morphological and life cycle characteristics.

TOPICS
1: PHYLUM CONIFEROPHYTA. GYMNOSPERMS
   A. MALE CONES
   B. FEMALE CONES
   C. SEEDS
2: PHYLUM ANTHOPHYTA. ANGIOSPERMS
   A. FLOWER PARTS
   B. DEVELOPMENT OF THE MALE GAMETOPHYTE AND POLLEN TUBE FORMATION
   C. DEVELOPMENT OF THE FEMALE GAMETOPHYTE AND FERTILIZATION
   D. SEEDS
3: CONSTRUCTION OF A PHYLOGRAM

INTRODUCTION
   In this exercise you will study gymnosperms and angiosperms, the seed plants. Seeds produced by both groups are advantageous for life on land because they protect the embryo from desiccation. Gymnosperms produce uncovered seeds. Pine trees and other conifers are the most common gymnosperms. All plants that have flowers are angiosperms. Angiosperms produce covered seeds. For example, bean seeds are enclosed in a pod that originated from female flower structures. The angiosperms include many trees and flower-producing plants.

Gymnosperms and angiosperms have other advanced traits in common:
- The sporophyte generation is dominant in gymnosperms and angiosperms.
- Gymnosperms and angiosperms are functionally heterosporous. They produce two types of spores, *megaspores* and *microspores* (Figure 1).
- Male and female gametophytes are markedly reduced.
- Well-developed vascular tissue enables gymnosperms and angiosperms to live in dry habitats.
- Vascular tissue functions to provide structural support for the plant.

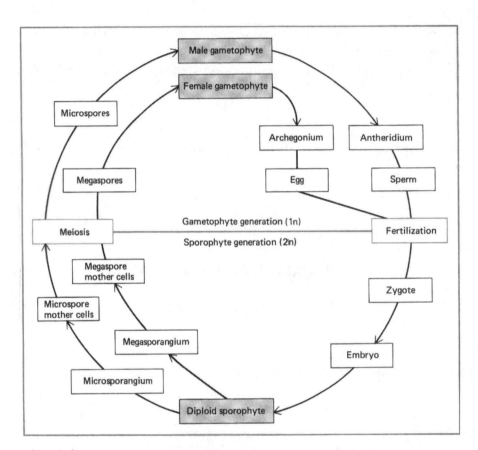

Figure 1. A heterosporous plant life cycle; the plants produce two types of spores, megaspores and microspores

1: PHYLUM CONIFEROPHYTA. GYMNOSPERMS

Pinus, the pine genus, is a typical gymnosperm. The tree is a sporophyte and is the dominant phase of the life cycle. Pine trees have well-developed xylem and phloem and can grow to be very large. The gametophyte phase is greatly reduced. The male gametophyte is composed of a small, multinucleate pollen grain. The female gametophyte is small and retained in the sporangium. In pine trees both male and female reproductive structures are produced on the same plant.

Several traits seen in pines and other gymnosperms eliminate the dependence on an aquatic habitat for sexual reproduction. Pollen is resistant to drying. The outer wall of the pollen grain has projections to facilitate wind dispersal. After wind carries a pollen grain to a female cone, a pollen tube digests its way to the egg and carries the sperm nucleus to the egg nucleus. After fertilization the zygote remains embedded in female gametophyte tissue, which nourishes the developing embryo. Gymnosperm seeds are "naked." Seeds are not covered by fruits as they are in angiosperms. Instead the embryo remains embedded in the megasporangium.

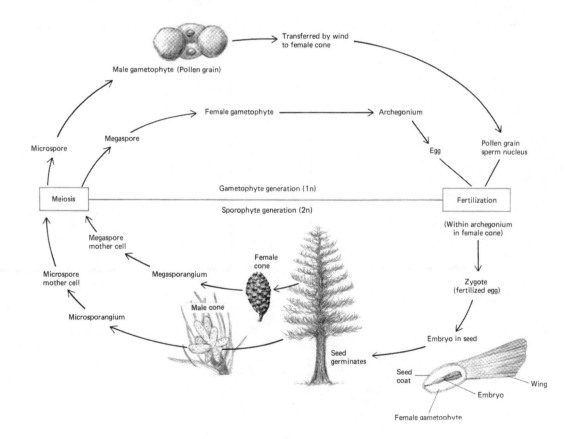

Figure 2. Gymnosperm life cycle

A. MALE CONES

Cones are reproductive structures that are composed of modified leaves called *sporophylls*. The sporophylls are arranged in a whorl around a central axis to form the cone. The male cones cluster at the tips of low branches. On the surface of each sporophyll of the cone are two microsporangia. Microspores are produced by meiosis in the microsporangia. The haploid nucleus of a microspore undergoes mitotic divisions to produce a pollen grain, the male gametophyte. Pollen grains are released from male cones in great abundance in the spring.

PROCEDURE

1. Examine a dried branch with male cones attached.
2. Use a microscope on low power to examine a prepared slide of a longitudinal section of a male cone.
3. Draw a microsporangium on a sporophyll in the space provided. Look for pollen inside the microsporangium. Examine a prepared slide of pine pollen, the male gametophyte. The "wings" are a modification of the wall of the pollen grain and aid in wind dispersal. Draw what you see in the space provided.

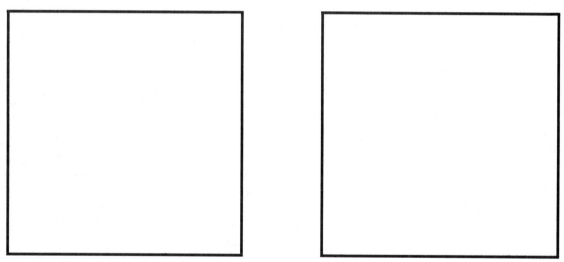

Pinus male cone longitudinal section   *Pinus* pollen grains

B. FEMALE CONES

The female cone is also composed of sporophylls arranged in a whorl around a central axis. Female cones are borne at the tip of short lateral branches in the upper portion of a pine tree and are larger than male cones from the same tree. On the upper surface of each sporophyll are two megasporangia. A megasporangium and its covering are called an *ovule*. A cell in the diploid megasporangium will undergo meiosis to produce four haploid megaspores. Three degenerate and the fourth undergoes a series of mitotic divisions to form the female gametophyte. Some gametophyte cells form two archegonia. Each archegonium produces an egg. The remaining cells of the female gametophyte provide nourishment for the embryo.

The first year female cones are small and soft in texture. Because of their location at the tip of a branch, young cones are not clearly visible. At about the same time pollen is released, the young female cones open, allowing the pollen to enter. After pollination, the female cone begins to harden. The pollen grain produces a slender pollen tube that digests its way to the egg. A sperm nucleus fertilizes the egg to produce a diploid zygote. Because the growth of the pollen tube and the development of the female gametophyte are quite slow, fertilization may not occur for as long as thirteen months following pollination. A hard female cone with mature seeds is about two years old in the southern United States and three years old in colder climates.

PROCEDURE

1. Examine a branch with female cones attached. How do they compare with male cones on the basis of size, shape and location on the tree?
2. Use microscope on low power to examine a prepared slide of a longitudinal section of a female cone.
3. Draw and label an ovule in the space provided.

Pinus female cone longitudinal section

C. SEEDS

After fertilization the diploid zygote divides to produce an embryo that is embedded in nutrient-rich haploid female gametophyte cells. The embryo digests and absorbs the gametophyte tissue. The diploid embryo and haploid female gametophyte are surrounded by coverings. One layer of the covering becomes hard and serves as the seed coat. A seed is a fertilized ovule in a protective seed coat.

As the embryo grows, seed leaves called _cotyledons_ form. Before germination of the seed, the cotyledons absorb nutrients from gametophyte tissue. After germination of the seed, cotyledons will begin to photosynthesize to provide food for the young seedling.

PROCEDURE

1. Cut through the female gametophyte tissue to locate the embryo.
2. Observe the gametophyte tissue and the embryo with a dissecting scope.
3. Draw what you see in the space provided. Label the gametophyte tissue and the embryo.

Pinus embryo

2: PHYLUM ANTHOPHYTA. ANGIOSPERMS

Angiosperms are the most recent and successful group of plants on earth. The group is diverse in terms of both habitat and morphology. Today you will see the basic reproductive characteristics of this group.

A. FLOWER PARTS

A flower is a cluster of leaf-like structures modified for reproduction. An almost infinite variety of sizes and shapes of flowers can be found in angiosperms.

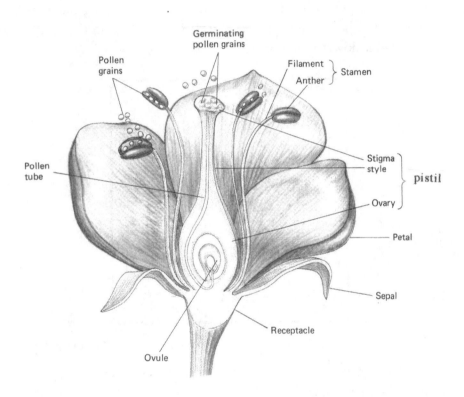

Figure 4. A typical flower

PROCEDURE
1. Dissect a flower.
2. Locate and know the parts listed below:
   Sepals: the protective outer whorl of green floral leaves.
   Petals: the whorl of floral leaves above the sepals.
   Stamens: another whorl of structures enclosed within the petals. Stamens are the male part of the plant. The stamen consists of a lower stalk called the *filament* and an upper enlarged portion called the *anther*. The anther contains the microsporangia that produce microspores. Microspores become pollen grains.
   Pistil: the structure in the center of a flower containing the female reproductive structures. The enlarged basal portion is the ovary containing the ovules where the eggs develop after pollination. The neck is the style. The stigma is at the top.
3. Draw and label a flower in the space provided.

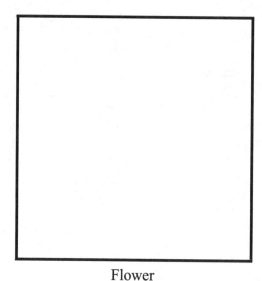

Flower

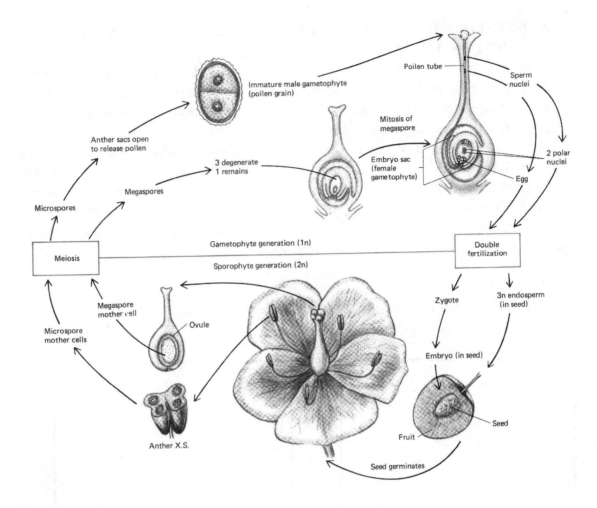

Figure 5. Angiosperm life cycle

B. DEVELOPMENT OF THE MALE GAMETOPHYTE AND POLLEN TUBE FORMATION

Meiosis occurs in the anthers to produce a tetrad of haploid microspores. Each microspore nucleus will undergo mitosis to form a cell with two nuclei. Then one nucleus undergoes an additional division to produce a pollen grain with three nuclei. A pollen grain with three nuclei is the mature male gametophyte.

Pollen is dispersed from the anthers. The grains may be blown about by wind or carried by insects. Pollen grains land on the stigma of a flower of the same species. The pollen grain then germinates, forming a tube that digests its way down the style and into an ovule. One pollen grain nucleus directs the synthesis of extracellular enzymes that digest the female tissues as the pollen tube grows. The other two nuclei are sperm nuclei. Each of the sperm nuclei is involved in fertilization.

PROCEDURE

1. Examine a prepared slide of pollen grains with germinating pollen tubes.
2. Draw in the space provided.

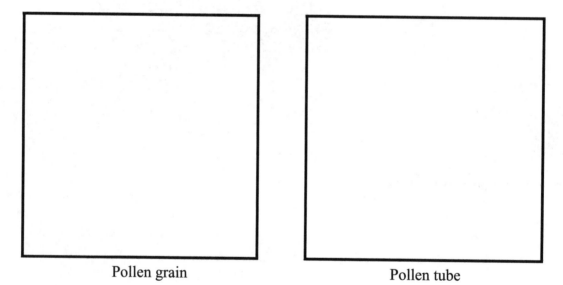

Pollen grain   Pollen tube

C. DEVELOPMENT OF THE FEMALE GAMETOPHYTE AND FERTILIZATION

Eggs are produced in ovules that are located inside the ovary. Each ovule has a large diploid megaspore mother cell that undergoes meiosis to produce four haploid megaspores. Three of the megaspores degenerate. The surviving haploid megaspore begins a series of three mitotic divisions that result in the formation of an embryo sac with eight haploid nuclei: the egg and seven other nuclei. The egg is the female gamete and fuses with a sperm nucleus from the pollen tube to form a diploid zygote. The zygote gives rise to the embryo. Two of the remaining seven nuclei are polar nuclei. The other sperm nucleus from the pollen tube fuses with the two polar nuclei to form a triploid endosperm nucleus. The endosperm nucleus will divide repeatedly to form endosperm tissue that will nourish the developing embryo. Double fertilization is a term that refers to the two separate fertilization events. One fertilization is the fusion of egg and sperm nuclei to form a diploid zygote. The other fertilization is the fusion of two haploid polar nuclei with a sperm nucleus to form a triploid endosperm. The other five haploid nuclei in the embryo sac disintegrate.

PROCEDURE
1. Study the model series of embryo sac development.
2. Draw a mature embryo sac in the space provided. Label the egg nucleus and the polar nuclei.

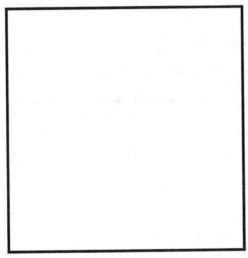

Mature embryo sac

3. What is formed when the egg is fertilized by a sperm? What is the ploidy of this tissue?
4. What is formed when the polar nuclei are fertilized by a sperm? What is the ploidy of this tissue?

D. SEEDS

Following fertilization, the diploid zygote divides to produce an embryo that is nourished by the triploid endosperm. In many plants the endosperm is fully used by the embryo. Tissue layers on the outside of the seed grow to harden to form the seed coats. A seed is a fertilized ovule in a protective seed coat. The ovary remains on the parent and often enlarges to form a fruit. The embryo consists of a stem apex that will grow into the stem, a root apex that will grow into the root, and one or two cotyledon that will serve as embryonic leaves to nourish the young plant during germination.

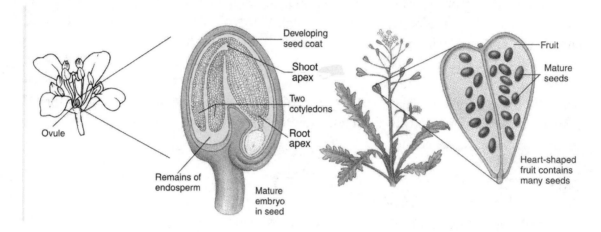

Figure 6. Embryonic development in the angiosperm seed

PROCEDURE

1. Examine a prepared slide of a *Capsella* embryo. Is any endosperm remaining?
2. Draw and label the cotyledons, shoot apex and root apex.

Capsella embryo

3. Examine a bean pod. What has happened to the pistil? Which structures have withered and fallen off and which remain? Open the pod and note where the seeds are located. How does their location compare to the position of the ovules in the young ovary?

4. Use a razor blade to make a longitudinal section of one of the seeds. Note the embryo with its large cotyledons. Draw a bean seed. Label the cotyledons, the stem apex and the root apex.

5. Draw a bean pod and label the stigma, style and ovaries.

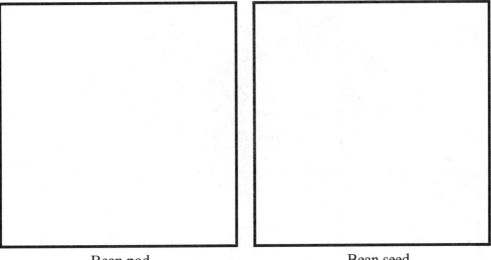

Bean pod   Bean seed

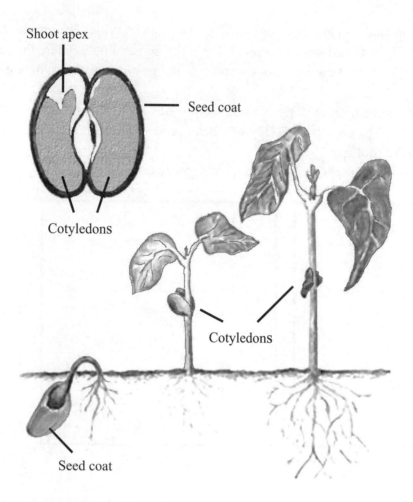

Shoot apex

Seed coat

Cotyledons

Cotyledons

Seed coat

Figure 7. Seed germination and growth of a young bean plant. As the seedling emerges from the ground, the seed coat is shed. The cotyledons photosynthesize and nourish the young seedling. The stem apex in the seed becomes the shoot with its first true foliage leaves. The root apex develops into the root. The growing leaves nourish the plant, and the cotyledons shrivel and fall off the stem.

3: CONSTRUCTION OF A PHYLOGRAM

PROCEDURE

1. Complete Table 1 by placing a "+" or a "0" in the blank section. If the phylum exhibits the characteristic in question, place a "+" in the blank space. If the characteristic is not present in the phylum, place a "0" in the blank.

Table 1. Summary of characteristics of plant phyla

Characteristics			
Phylum	Sporophyte dominant (+ or 0)	Seed producer (+ or 0)	Covered seed (+ or 0)
1. Bryophyta			
2. Pterophyta			
3. Coniferophyta			
4. Anthophyta			

2. Compare the plant phyla to one another and indicate the number of "+" characteristics between each pair. Complete Table 2.

Table 2.   Total number of shared characteristics

Total number of shared characteristics		
1, 2 =	2, 3 =	3, 4 =
1, 3 =	2, 4 =	
1, 4 =		

3. Construct a phylogram of plant phyla.

Exercise 17

4. Using the information in Table 1, complete Table 3.

Table 3. Summary of characteristics of plant phyla

Characteristics		Phylum
Dominant phase of the life cycle	Gametophyte dominant	
	Sporophyte dominant	
Seeds	Not a seed producer	
	Seed producer	
Covered seeds	Seeds not covered	
	Seeds covered	

5. The phylogram in Table 4 shows evolutionary relationships in the plant kingdom. The completed boxes on the tree are characteristics that are used to separate groups of plants from one another. Write the correct plant phylum in the empty boxes.

Table 4. Phylogram of the plant kingdom

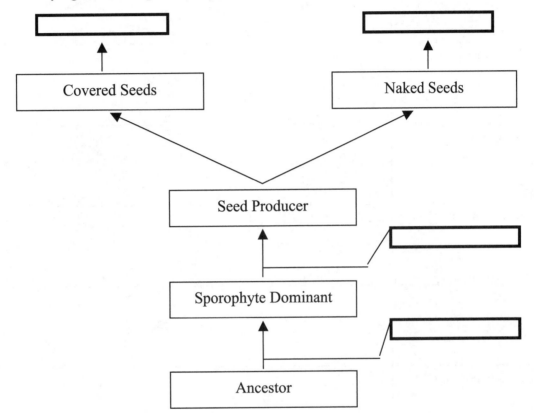

VOCABULARY

TERM	DEFINITION
Angiosperm	
Gymnosperm	
Megaspore	
Microspore	
Cones	
Microsporangium	
Megasporangium	
Ovule	
Flower	
Sepal	

TERM	DEFINITION
Stamen	
Filament	
Anther	
Pistil	
Stigma	
Style	
Ovary	
Megaspore mother cell	
Embryo sac	

TERM	DEFINITION
Polar nuclei	
Sperm nuclei	
Endosperm	
Double fertilization	
Cotyledons	
Shoot apex	
Root apex	

Exercise 17

QUESTIONS

1. Why is the development of a seed an advantage for life on land?

2. What is the function of the wings on a pollen grain?

3. Describe the male and female gametophytes in gymnosperms.
   Male:

   Female:

4. What is the ploidy of tissue that nourishes a gymnosperm embryo?

5. Describe the male and female gametophytes in angiosperms.
   Male:

   Female:

6. Review the formation of a mature embryo sac.

7. Describe double fertilization. What is the ploidy of the tissue produced by double fertilization that nourishes an angiosperm embryo?

8. What happens to the young cotyledons as a young seedling grows?

9. Review the characteristics used to construct a phylogram of plant groups.

KINGDOM ANIMALIA
PHYLA PORIFERA, CNIDARIA AND PLATYHELMINTHES

OBJECTIVES
- Define the vocabulary terms from the questions at the end of this exercise.
- Recognize the major characteristics of the Porifera, Cnidaria and Platyhelminthes phyla.
- Understand how the major characteristics suggest phylogenetic relationships between the phyla.

TOPICS
1: PHYLUM PORIFERA. SPONGES
2: PHYLUM CNIDARIA.
   HYDRA, *OBELIA*, JELLYFISH, SEA ANEMONES, CORAL
3: PHYLUM PLATYHELMINTHES. FLATWORMS

INTRODUCTION
Animals are multicellular, heterotrophic organisms. They cannot make their own food from inorganic molecules and must obtain these organic molecules from the outside. Most biologists agree that animals evolved from protists, but they are not sure exactly which protist line. The most primitive animals are the sponges. They have a simple body plan and lack symmetry and true tissues. Only the Cnidarians have radial symmetry throughout their life cycle. All other phyla have bilateral symmetry at some point in their life cycle. The bilaterally symmetrical phyla are further divided according to type of body cavity, the space between the body wall and the digestive tract. Acoelomate organisms lack a coelom, a fluid-filled body cavity that is completely surrounded with tissues of mesodermal origin; pseudocoelomate organisms have a body cavity that is not completely surrounded by mesodermal tissues; and coelomate organisms have a true coelom.

Pseudocoelomate and coelomate animals have a complete digestive tract with a mouth and an anus. However, they differ in the timing of formation of the mouth and anus. In early embryonic development an opening called a blastopore develops in the primitive blastula. The blastopore marks the beginning of the gut and will become either a mouth or an anus. In protostomes, "mouth first," the blastopore develops into a mouth, and the anus is formed later. In deuterostomes, "mouth second," the blastopore develops into the anus, and the mouth develops later.

In this exercise and the next two exercises you will study representatives of nine animal phyla. You will learn distinguishing characteristics of each phylum so that you can make comparisons between them. You will use these characteristics to construct a phylogenetic tree that will indicate evolutionary relationships among the animals you will study.

246

The phylogenetic tree will be based on five major characteristics:

- Level of organization. Are true tissues present?

- Symmetry. Is the animal radially or bilaterally symmetrical?

- Body cavity. What kind of body cavity is present? Bilaterally symmetrical animals are distinguished by one of three kinds of body cavities:

    a. Acoelomate—No coelom (body cavity). The gut lies in a space entirely filled with tissue of mesodermal origin.
    b. Pseudocoelomate—False coelom. The body cavity is partially surrounded by tissue of mesodermal origin.
    c. Coelomate—True coelom. The body cavity is completely surrounded by tissue of mesodermal origin.

- Fate of the blastopore. Is the animal a protostome or a deuterostome? The blastopore is the first opening in the embryonic blastula. If the blastopore becomes the mouth, the animal is a protostome. If the blastopore becomes the anus, the animal is a deuterostome.

- Segmentation. Is the body composed of repeated segments?

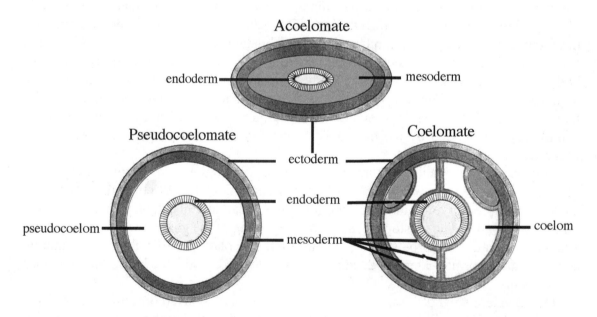

Figure 1. Acoelomate, pseudocoelomate and coelomate body plans

As you study the animals in Exercises 9, 10 and 11, you will record their characteristics in the table at the end of Exercise 11. You will use these characteristics to make a phylogram that will reflect the evolutionary relationships among them. In this exercise you will study the most primitive animals: the sponges, the Cnidarians and the Platyhelminthes. You will study the protostomes in Exercise 10 and the deuterostomes in Exercise 11.

1: PHYLUM PORIFERA

Sponges have a very primitive level of body organization. There is no symmetry in most sponges and no organization of cells into tissues. Since sponges do not have a body cavity, they have no coelomic plan. A sponge is porous sac composed of two cell layers. The outer cell layer is composed of epidermal cells, and the inner layer contains flagellated collar cells. The beating flagella move water through pores into a large central cavity and out an opening called an osculum. Sponges have a distinctive way of feeding. Nutrients in the water are trapped in a sticky mucus at the base of the flagellum of a collar cell. They are digested in food vacuoles inside the collar cell and then passed to other cells. Many sponges have thin, pointed structures called spicules that give strength to the sponge body.

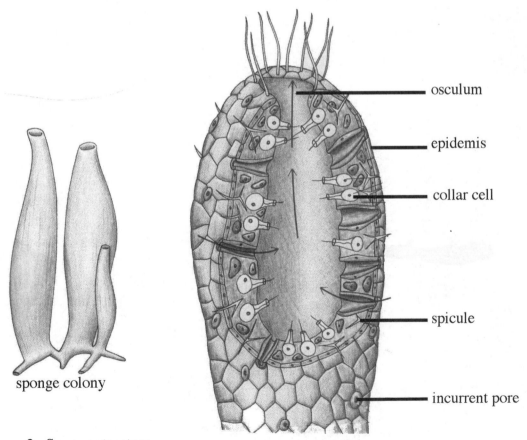

Figure 2. Sponge structure

PROCEDURE

1. Observe a whole sponge under a stereomicroscope. Note the osculum.

2. Examine a prepared slide of a sponge. Draw the sponge and label the osculum, central cavity, pores and spicules in the space provided.

Sponge longitudinal section

2: PHYLUM CNIDARIA

Cnidarians are more advanced than sponges. They have distinct tissues, symmetry and two tissue layers. The outer layer, the epidermis, covers the external surface of the body. The inner layer, the gastrodermis, lines an unbranched gastrovascular cavity that has only one opening. Gelatinous mesoglea lies between the two germ layers. Cnidarians do not have a body cavity and thus are not considered to have any kind of coelomic plan. There are two basic body forms in cnidarians: the polyp stage that is usually attached to a substrate, and the medusa stage that is free-living. The polyp and medusa stages alternate in the life cycle. You will observe five members of the phylum Cnidaria. Note that cnidarians have radial symmetry; body parts are arranged around a central axis.

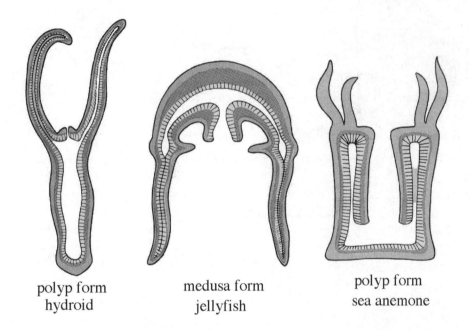

polyp form
hydroid

medusa form
jellyfish

polyp form
sea anemone

Figure 3. The polyp and medusa forms of Cnidarians are similar structurally. (a) The polyp form in *Hydra*. (b) The medusa form is a polyp turned upside-down. (c) The polyp form in the sea anemone.

A. *Hydra*

Hydra exists only in a polyp stage. It has contractile fibers in the cells of both the epidermis and gastrodermis. In the epidermis they run longitudinally so that contraction results in shortening of the body. In the gastrodermis they run perpendicularly so their contraction results in decrease of the body diameter. A net of interconnecting nerve cells is located within the mesoglea. The nerve net and contractile fibers are responsible for the motor responses you will see in the living organism. The gastrovascular cavity is unbranched.

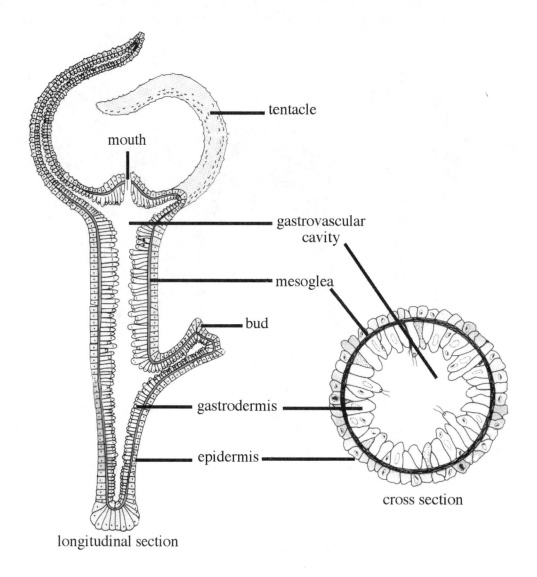

Figure 4. Hydra body structure

PROCEDURE

1. Examine the live *Hydra* under the stereomicroscope. Note the foot, tentacles and mouth. Gently touch one of the tentacles with a dissecting needle and note what happens.

2. Examine a whole mount prepared slide of Hydra. Draw and label the tentacles and mouth in the space provided.

Hydra

B. *Obelia*

Obelia has a life cycle of alternating polyp and medusa phases. The polyps reproduce asexually by budding. The fully developed buds usually remain attached to the colony. There are two kinds of polyps, feeding polyps and reproductive polyps. Feeding polyps have a mouth encircled by a ring of tentacles bearing stinging cells. Inside the reproductive polyps the small bell-shaped medusae develop. When mature, the medusae are released to live a free-swimming existence. The medusae produce gametes that unite to form a zygote. The zygote divides and differentiates to form a swimming larva that eventually develops into a polyp. What adaptive advantage is served by the motile medusa stage?

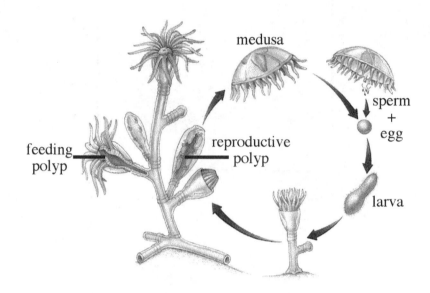

Figure 4. *Obelia* life cycle

PROCEDURE

1. Examine a prepared slide of *Obelia*.

2. Draw and label the feeding polyps, tentacles, reproductive polyps, young medusae and the gastrovascular cavity in the space provided.

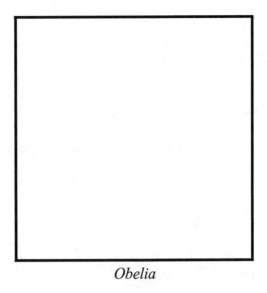

Obelia

C. Jellyfish

Jellyfish float near the ocean surface. The medusa stage is dominant.

<u>PROCEDURE</u>
1. Examine the preserved jellyfish. Note the tentacles and mouth.

2. Draw the jellyfish and label the tentacles and mouth in the space provided.

Jellyfish

D. Sea anemone

Sea anemones are the most advanced cnidarians. Their structure is more elaborate and complex than Hydra. Sea anemones are polyps. There is no medusa in the life cycle.

<u>PROCEDURE</u>
1. Examine the preserved sea anemone. Note the tentacles.

2. Draw the sea anemone and label the tentacles and mouth in the space provided.

Sea anemone

E. Coral

Like the sea anemone, coral does not have a medusa stage. Corals exist only as polyps. They feed on small prey mostly during the night. Most corals have a symbiotic relationship with green algae and other photosynthetic protists that live within the coral tissue. During the day, these photosynthesizers provide nutrients for the coral. The coral polyp secretes a calcium carbonate skeleton that forms coral reefs.

<u>PROCEDURE</u>
1. Examine a coral skeleton. Note the openings where the polyps lived.

3: PHYLUM PLATYHELMINTHES

The Platyhelminthes, or flatworms, show important structural advances over the previous phyla. They have three distinct tissue layers, epidermis derived from ectoderm, endodermis derived from endoderm and a middle tissue layer derived from mesoderm. Platyhelminthes are the first organisms that have a body cavity. Since the body cavity where the organs lie is filled with tissue of mesodermal origin, the Platyhelminthes are acoelomate. Platyhelminthes have bilateral symmetry that permits cephalization, the concentration of the nervous system in the head. Some flatworms are free-living like *Planaria*. Others are parasitic like the tapeworms.

A. *Planaria*

Planaria is a free-living flatworm. Some species are found in moist soil and others are found in water. *Planaria* has a gastrovascular cavity that lies in a body cavity filled with mesoderm. There is only one opening, the mouth. Food enters the mouth, moves into a large pharynx and then into the branching gastrovascular cavity. The gastrovascular cavity is more advanced than in the Cnidarians because the extensive branching allows greater surface area for absorption of nutrients. Undigested food is released from the mouth. The gastrovascular cavity has both digestive and circulatory functions. The eyespots are light receptors and the auricles, ear-like bulges at the side of the head, are chemoreceptors.

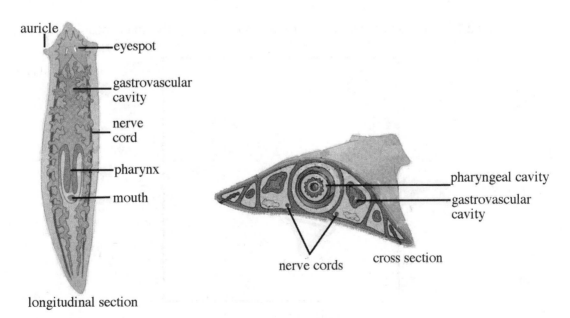

Figure 5. *Planaria*

256

PROCEDURE

1. Examine the live *Planaria* under the stereomicroscope.

2. Draw *Planaria* and label the eyespots in the space provided.

Live *Planaria*

3. Whole mount slide. Observe a prepared whole mount slide of a *Planaria* that has been injected to show its branching gastrovascular cavity. How is the gastrovascular cavity similar to the gastrovascular cavity of Hydra? How does it differ? Does it seem likely that *Planaria* would have a circulatory system?

4. Draw a *Planaria* and label the gastrovascular cavity in the space provided.

Whole mount *Planaria*

5. Cross section slide. On a preserved cross section slide of a *Planaria* find the epidermis. In the middle section of the *Planaria* note the large pharynx and the smaller branches of the gastrovascular cavity. Note that the organs lie in a space that is filled with solid material. This is mesodermal in origin. Does *Planaria* have a coelom?

6. Draw the *Planaria* cross section and label the pharynx and branches of the gastrovascular cavity in the space provided.

Planaria cross section

B. Tapeworm

Tapeworms are parasitic flatworms that have no digestive system of their own. They attach to the host digestive tract by hooks and suckers that are located on the head region called the scolex. They absorb nutrients through their body surface. The body surface is similar to the lining of the host digestive tract and is so well adapted for absorption of nutrients that it has villi and microvilli, like the host. Proglottids are continually formed behind the scolex. Each proglottid develops both male and female organs as it matures. Mating takes place inside the proglottid. In mature proglottids, at the far end of the tapeworm, eggs hatch into larvae. The last proglottids bulge with hundreds of larvae. These proglottids break off and are shed through the feces. Although proglottids appear to be segments, they are not true segments. Like all other parasitic flatworms, the tapeworm life cycle has two hosts. An intermediate host eats the larvae. The larvae bore through the digestive tract of the host and travel via the bloodstream to muscle where they form cysts. The primary host eats the muscle, and the tapeworm leaves the cyst and attaches to the intestinal wall where it completes the sexual cycle.

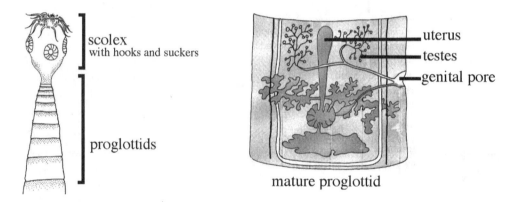

Figure 6.   Tapeworm

<u>PROCEDURE</u>
1. Examine the demonstration of a preserved tapeworm. Identify the scolex, hooks and suckers, and proglottids. Note the large, mature proglottids at the far end of the tapeworm.

2. Draw a tapeworm and label the scolex, hooks and suckers, and proglottids in the space provided.

Exercise 18

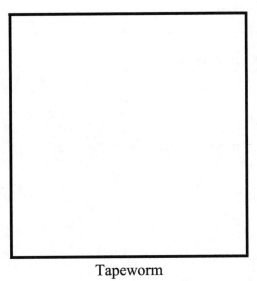

Tapeworm

VOCABULARY

TERMS	DEFINITION
Heterotrophic	
Radial symmetry	
Bilateral symmetry	
Coelom	
Acoelomate	
Pseudocoelomate	
Coelomate	
Collar cells	
Osculum	
Spicules	

TERMS	DEFINITION
Epidermis	
Gastrodermis	
Mesoglea	
Gastrovascular cavity	
Polyp	
Medusa	
Ectoderm	
Endoderm	
Mesoderm	
Cephalization	
Scolex	
Proglottid	

QUESTIONS

1. Why are sponges considered to be primitive animals?

2. How does a sponge obtain food?

3. Name three advances that cnidarians have over sponges.

4. What phylum has members that have radial symmetry?

5. Describe the life cycle of *Obelia*. What adaptive advantage is served by the motile medusa stage?

6. Explain the symbiotic relationship found in corals.

7. Name three advances that flatworms have over cnidarians.

8. Explain how the development of bilateral symmetry allows for cephalization.

264

9. Compare the *Hydra* and *Planaria* gastrovascular cavities. Which has the more advanced gastrovascular cavity? Why?

10. Do *Hydra* and *Planaria* need a circulatory system? Why?

11. Record the characteristics of the phyla in this exercise in Table 1 of Exercise 11.

EXERCISE 19

KINGDOM ANIMALIA
PROTOSTOMES
PHYLA NEMATODA, MOLLUSCA, ANNELIDA, ARTHROPODA

OBJECTIVES
- Define the vocabulary terms from the questions at the end of this exercise.
- Know major characteristics of the Nematoda, Mollusca, Annelida and Arthropoda.
- Understand how the major characteristics suggest phylogenetic relationships between the phyla.

TOPICS
1:    PHYLUM NEMATODA. ROUNDWORMS
2:    PHYLUM MOLLUSCA. CLAM
3:    PHYLUM ANNELIDA. EARTHWORM
4:    PHYLUM ARTHROPODA.   TRILOBITE AND GRASSHOPPER

INTRODUCTION
       The animals in this exercise are the first that have a complete digestive tract from mouth to anus. They are protostomes because the mouth develops from the blastopore and the anus is formed later. In this exercise you will study representatives of five protostome phyla: Nematoda, Mollusca, Annelida and Arthropoda. Nematodes have a pseudocoelom. Molluscs, annelids, trilobites and arthropods have a true coelom.

1:   PHYLUM NEMATODA. ROUNDWORMS

       Nematodes or roundworms are unsegmented roundworms.   They are pseudocoeloms because their body cavity is not completely surrounded by mesoderm. The development of a fluid-filled body cavity is a great evolutionary advance because it provides space in which organs can lie. Nematodes have a complete digestive tract with a mouth and an anus. As in other protostomes, the mouth develops before the anus. Some nematodes are free living and may be present in very large numbers. A spade of soil or a bucket of water can contain more than one million nematodes. Other nematodes are parasitic on either plants or animals.

266

A. *Ascaris*

Ascaris is parasitic in many mammals including humans, cats and dogs. Over one billion humans in the world are infected with *Ascaris*. Adult worms attach to the small intestine where they feed on nutrients. *Ascaris* is covered by a thick cuticle that resists digestion by the host. The sexes are separate. After mating, the female releases more than a million fertilized eggs into the host intestine where they are shed in the feces. Larvae begin development in a suitable habitat, like moist soil. The developing larvae are eaten and then hatch in the intestine of the host. Before they can complete development to the adult stage, they begin a long migration through the intestinal wall to the blood stream, and eventually to the lungs. They travel from the lungs up the bronchi to the pharynx and then down to the esophagus, stomach and intestine again. Once back in the intestine, they complete development to sexually mature adults, thus completing the cycle.

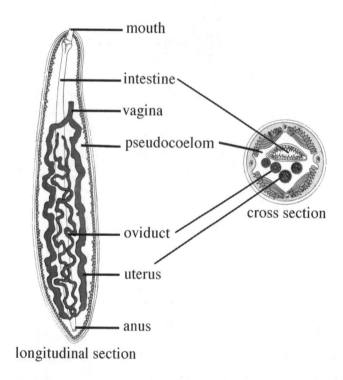

Figure 1.   *Ascaris* female longitudinal and cross sections

PROCEDURE

1. Examine the preserved dissection of *Ascaris* on display in the lab. The body cavity is evident.

2. Study the external features of a preserved specimen of *Ascaris*. Note the round body shape and thick cuticle that resists digestion by the host.

3. Locate the mouth and the anus.

4. Determine the dorsal surface by locating the anus that is on the ventral side of the body.

5. Cut the specimen along the midline of the dorsal surface. Pin the body wall back so that the organs are exposed. Note the pseudocoelom, the body cavity in which organs lie. The tissue lining the outside of the cavity is muscle that is derived from mesoderm.

6. Separate the digestive tract from the other organs. Note the thin-walled intestine. It is a nonmuscular tube formed from tissue of endodermal origin. Locate the ovaries, oviducts, and uterus in the female or the seminal vesicles and testes in the male.

7. Draw an *Ascaris* in the space provided. Label the mouth, anus, intestine, and in a female, the ovaries, oviducts, and uterus or in a male, the seminal vesicles and testes.

*Ascari*s

B. Dog heartworm

Unlike *Ascaris* where there is only one host necessary for adults to complete a life cycle, the dog heartworm life cycle requires two hosts, a mosquito and a dog. Mosquitoes suck the blood of an infected dog. Larvae develop in the mosquito. The mosquito bites another dog, and the larvae enter muscle where they develop into immature adult worms. The worms enter the blood stream and move to the heart where they mature and mate. Large numbers of worms clog the right heart and pulmonary arteries causing severe illness. Examine the infected dog heart.

2:   PHYLUM MOLLUSCA. CLAM

Molluscs have a true coelom. The coelom is reduced to small cavities in which the heart, reproductive and excretory organs lie. Most molluscs are found in marine habitats, but there are also freshwater and terrestrial representatives. Included in this phylum are the chitons, oysters, clams, snails, slugs, squids, nautili and octopods. Soft unsegmented bodies characterize this phylum.

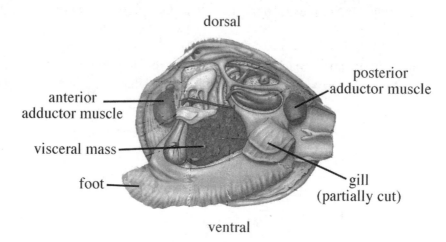

Figure 2. Clam

PROCEDURE

1. Study the external features of the clam.

2. Balance the clam on the open edge of its shell. The uppermost surface is the dorsal surface. The lower surface is the ventral surface. The bulge on the dorsal surface is the umbo. The anterior end, where the head is located, is rounded. The posterior end is pointed.

3. Look into the clam and use a scalpel to cut the anterior and posterior adductor muscles that contract to close the shell.

4. Carefully open the clam. Next to the shell is a thin tissue layer called the mantle. The mantle secretes a calcareous shell. In some more specialized molluscs the shell has become reduced, either embedded in the soft tissue or lost altogether.

5. Cut away the mantle to expose the mantle cavity. The gills are located in this cavity. The beating of cilia on the gills forces water across the gills where gas exchange takes place. Clams are filter feeders. They filter small food particles through the pores in their gills as water passes through.

6. The main body of the clam consists of the muscular foot and the visceral mass dorsal to the foot. The foot is used for locomotion. The visceral mass contains the digestive, excretory and reproductive organs.

7. Draw a clam in the space provided. Label the umbo, anterior and posterior adductor muscles, mantle, foot, visceral mass and gills.

Clam

3:	PHYLUM ANNELIDA. EARTHWORM

Annelids are segmented. In addition to segmentation, annelids exhibit other important structural advances over the flatworms and roundworms. Among these advances are a closed circulatory system, a central nervous system and a true coelom. The fluid in the coelom serves as a hydrostatic skeleton.

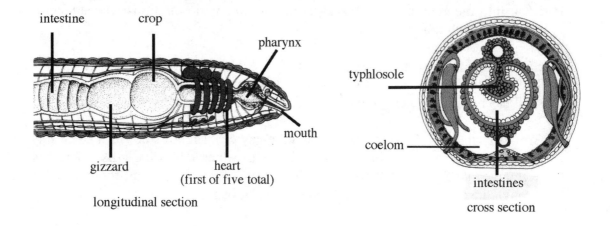

Figure 3. Earthworm

PROCEDURE

1. Study the external features of the earthworm. Note the segmentation. Each segment is a metamere.

2. Pass your fingers up and down the sides of the worm. On the ventral side you can feel bristles called setae. The enlarged, light segment is the clitellum. The opening near the clitellum is the mouth. The opening at the other end is the anus.

3. Place the worm on its ventral surface in the dissecting pan. Use a scalpel to make a shallow incision slightly to one side of the dark blood vessel beginning at the anus. Use a pair of teasing needles and scissors to carefully work your way toward the mouth without destroying the internal structures. Pin both sides every 10th segment.

4. Each earthworm has complete male and female sex organs and is therefore hermaphroditic. Even though one worm produces both male and female gametes, self-fertilization does not occur. The clitellum produces mucus that holds two worms together during copulation.

5. Count the number of segments. On the 15th segment you will find the openings of the seminal vesicles. The worm produces and stores sperm in the seminal vesicles. Sperm are released from seminal vesicles of one worm into seminal receptacles of its partner. After the worms separate, the clitellum secretes another mucus casing that passes toward the head and picks up first the eggs that were made by the worm and then the sperm which were deposited by the partner. The casing slips over the head and fertilization and hatching of the young take place inside.

6. The digestive tract is composed of muscle that is mesodermal in origin. Food passes from the mouth into a muscular pharynx located in segments 3 to 6, then passes down to the 14th segments to a crop for temporary storage. The gizzard in segments 17 and 18 grinds food with the aid of sand grains. Then food passes through the intestines and out the anus.

7. Find the mouth, pharynx, crop and gizzard. Why is the digestive tract muscular?

8. The circulatory system of the worm is composed of closed vessels. The major vessels are a dorsal vessel lying on top of the digestive tract and a ventral vessel below it. These two vessels are connected by five hearts located above the esophagus at the 7th through 11th segments. The hearts help to move blood through the vessels. Find the dorsal and ventral vessels and the five hearts.

9. Earthworms have a well developed nervous system with a two lobed brain located on the dorsal surface of the pharynx in segment 3. A nerve runs on either side of the pharynx connecting the brain to a pair of ganglia that lie below the pharynx in segment 4. Thereafter, each segment contains a pair of ganglia that control movement. A pair of ventral nerve chords running the length of the earthworm connects the ganglia. Find the brain.

10. Draw the earthworm in the space provided. Label the metameres, clitellum, setae, mouth, pharynx, crop, gizzard, intestines, anus, seminal vesicles, seminal receptacles, dorsal and ventral vessels, hearts and brain. Use a dissecting scope to study a prepared cross section slide of the earthworm. Note the thick outer tissue composing the body wall, the coelom and the inner circle of the intestinal wall. The intestine has a modification called the typhlosole that increases surface area to improve the absorption of nutrients.

Earthworm dissection

11. Draw the cross section of the earthworm in the space provided. Label the cuticle, coelom, intestine and typhlosole.

12. Review the earthworm structures by studying the model on the side bench.

Earthworm cross section

4:   PHYLUM ARTHROPODA

Along the side bench you will find examples of this large and diverse phylum. Phylum Arthropoda is the largest and most successful of the animal phyla with members in almost every habitat. Their bodies are covered with a hard exoskeleton made of chitin. Bodies are segmented internally and externally to various degrees. A head, thorax and abdomen are visible, though the head and thorax may be fused. Jointed appendages are one of the most apparent and versatile features of the arthropods. The nervous system is composed of a brain and a ventral nerve cord. The most conspicuous sensory organ is the pair of compound eyes. The circulatory system is open. The respiratory system has tremendous surface area that compensates for the inefficiency of an open circulatory system. Many arthropods breathe through a branched tracheal system. Air enters from numerous spiracles, tiny pores on the main body. Oxygen diffuses directly into tissues as needed so the blood does not need to move very quickly or in a precise pathway. Like the molluscs, the coelom is greatly reduced. In arthropods it consists only of the cavity that contains reproductive and excretory organs.

A. Trilobite

Trilobites are early arthropods that were abundant 500 million years ago. They have been extinct for 250 million years. Three longitudinal lobes, a median lobe and two lateral lobes, give them their name. Trilobites exhibit primitive segmentation, because the segments lack specialization.

Figure 4. Trilobite

PROCEDURE

1. Examine the trilobite fossil. Distinguish median and lateral lobes.

2. Note little difference or variation in the segments.

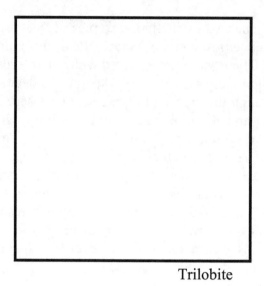

Trilobite

B. **Grasshopper**

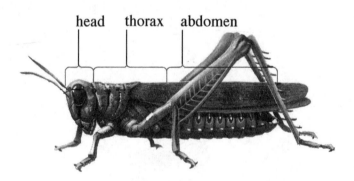

Figure 5. Grasshopper

PROCEDURE

1. Examine the preserved grasshopper. Distinguish the head, thorax and abdomen.

2. Observe the external sensory receptors. Antennae and compound eyes are easily visible.

3. Look at mouthparts with a dissecting scope.

4. Note the three pairs of legs. Each pair is attached to the thorax.

5. Females have a pointed ovipositor, which is used in digging a site for egg deposits. Males have a tapered abdominal end.

6. Locate the spiracles along the lateral sides of the thoracic and abdominal segments. These are the openings of the tracheal respiratory system.

7. Locate the two pairs of wings that are attached to the thoracic segment. Spread the wings and note the firm protective outer pair and the thinner inner pair.

8. Draw a grasshopper in the space provided. Label the head, thorax, abdomen, antennae, legs, wings and spiracles.

Grasshopper

276

VOCABULARY

TERM	DEFINITION
Cuticle	
Umbo	
Mantle	
Gills	
Foot	
Visceral mass	
Metamere	
Setae	
Clitellum	
Hermaphroditic	

Exercise 19

TERM	DEFINITION
Seminal vesicles	
Seminal receptacles	
Mouth	
Pharynx	
Crop	
Gizzard	
Intestine	
Anus	
Dorsal vessel	
Ventral vessel	
Heart	

TERM	DEFINITION
Typhlosole	
Exoskeleton	
Chitin	
Tracheal system	
Spiracles	
Head	
Thorax	
Abdomen	
Antennae	
Compound eyes	
Ovipositor	
Wings	

QUESTIONS

1. What advantage does the pseudocoelomate body cavity have over that of an acoelomate?

2. Which phylum that you studied is the first to have a complete digestive tract from mouth to anus?

3. What is the advantage of having a complete digestive tract from mouth to anus?

4. Compare the different methods of nutrient absorption in *Ascaris* and tapeworm. Why does *Ascaris* need a digestive tract?

280

5. *Ascaris* has a pseudocoelom because the digestive tract is composed of thin tissue of endodermal origin. Earthworms have a true coelom with a muscular digestive tract of mesodermal origin. Why does an earthworm require a muscular digestive tract when an *Ascaris* does not?

6. Which phyla in this exercise have segmentation?

7. What are the characteristics of primitive segmentation?

8. Record the characteristics of the phyla in this exercise in Table 1 of Exercise 11.

KINGDOM ANIMALIA
DEUTEROSTOMES
PHYLA ECHINODERMATA AND CHORDATA

OBJECTIVES
- Define the vocabulary terms from the questions at the end of this exercise.
- Recognize the major characteristics of the Echinodermata and Chordata phyla.
- Understand how the major characteristics suggest phylogenetic relationships between the phyla.

TOPICS
1: PHYLUM ECHINODERMATA. STARFISH
2: PHYLUM CHORDATA. RAT
3: CONSTRUCTION OF A PHYLOGRAM

INTRODUCTION

Besides being true coelomates, the Echinodermata and Chordata phyla share a major characteristic. They both are deuterostomes. The anus develops from the blastopore, and the mouth is formed later. You will study a representative of each phylum.

1: PHYLUM ECHINODERMATA. STARFISH

All members of the phylum Echinodermata are marine. The adults are radially symmetrical. Echinoderm ancestors were bilaterally symmetrical, a trait that is retained in the larval forms of present day echinoderms. They have a poorly developed circulatory system and no excretory system. Like most other radially symmetrical animals, there is no cephalization—no brain or head. However, there are rings of nerves around the central disc, and a nerve cord radiates out to each arm. Echinoderms have a unique water vascular system that is used in locomotion. They are unsegmented.

282

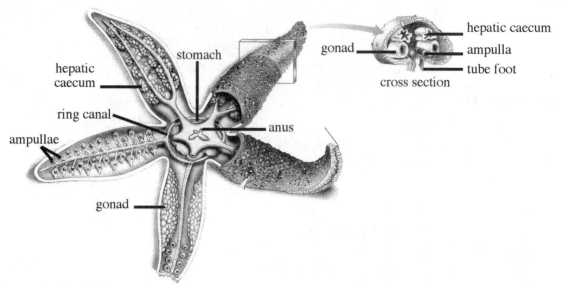

Figure 1. Starfish

PROCEDURE

1. Study the external features of the starfish. Locate the circular madreporite on the upper aboral surface. The anal opening is also on the aboral surface but it is difficult to find. The spines are part of the calcium plates that lie buried in the body wall beneath the surface.

2. On the lower oral surface locate the mouth protected by spines. The cardiac stomach may be visible protruding from the mouth. A groove studded with tube feet runs along each arm.

3. Place the starfish in a dissecting tray with the aboral surface facing up. Use dissecting scissors to cut off the tip of one arm. Then separate the entire aboral surface of the starfish from the oral surface by cutting along each arm toward the central disk.

4. Each arm contains a pair of digestive glands, the hepatic caecum. Under the hepatic caecum in each arm is a pair of feathery gonads, the reproductive organs. The sexes are separate in the starfish. Fertilization is external to the body. Locate the hepatic caecum and gonads.

5. The madreporite is the entrance to the water vascular system. Water enters through the madreporite, then goes into the stone canal and then into the ring canal. From the ring canal water flows down each arm via radial canals. Remove the hepatic caecum and gonads to view the stone canal, ring canal and radial canals. Find the stone canal and ring canal at the center of the starfish. The radial canal extends down each arm.

6. Branching from each radial canal are hundreds of tube feet. Attached to each tube foot is an ampulla, a bulb-shaped structure that contracts, pushing water into the tube foot and causing it to extend. When the tube foot extends, its suction cup can grip the surface. The nervous system of the starfish controls the flow of water into the ampulla. The starfish can attach the suction cups of its tube feet to the shells of bivalve molluscs and exert a steady pull that opens the mollusc so it can be eaten. The starfish can also move slowly by this method.

2: PHYLUM CHORDATA. RAT

Sometime in their lives members of the phylum Chordata possess a notochord, a dorsal hollow nerve chord, gill slits and a post-anal tail. You will study a typical chordate, the rat. The rat is very much like the human.

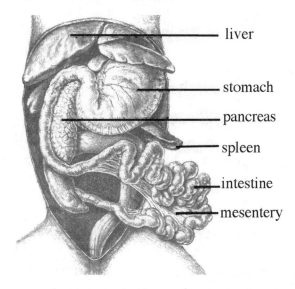

Figure 2. Rat

PROCEDURE
1. Begin by locating these external features.
Head—with a concentration of sensory receptors (eyes, ears, nostrils, mouth)
Trunk—including two pairs of appendages and a tail, the thorax, abdomen and anus

2. After making the external identifications, you are ready to begin the dissection. Place the animal on its back, the dorsal surface, in the dissecting tray and use pins through the four limbs to hold it down.

3. Pull up loose skin and muscle of the abdomen with forceps and snip a small hole with scissors. Insert the scissors so you can make a cut up to the throat, then cut down just short of the reproductive structures. Make a transverse cut in the skin and muscle at the bottom of the abdomen and the top of the thorax so you can pin it back out of the way. Note the membrane covering the viscera. The true coelom is inside the membrane.

4. Beginning in the neck and upper thorax, locate the large pink thymus, a gland important in the immune system of the rat.

5. Carefully cut through the ribs to expose the thoracic cavity. Locate the diaphragm that separates the thoracic cavity from the abdominal cavity. The diaphragm is a sheet of muscle that contracts and relaxes during breathing. Find the heart. Note the trachea that carries air to the lungs. The trachea divides into a right and left bronchus, which divides a number of times into the spongy lungs.

6. Make a longitudinal cut in the membrane covering the abdominal viscera to expose the viscera. Note the large dark red liver divided into three lobes. A small green sac on the underside of the right liver lobe is the gall bladder. The gall bladder stores bile and releases it into the upper small intestine.

7. Find the stomach, small intestine and large intestine. Notice as you uncoil the intestines that they are held together by a thin membrane, the mesentery. The pancreas is a diffuse, pink tissue located in the mesentery behind the stomach and upper small intestine.

8. Underneath the liver is the spleen. The spleen is darker than the liver and can be distinguished by its long thin shape. It is found to the left side of the stomach.

9. Under the intestines are the kidneys, small bean-shaped structures. They produce urine through filtration of the blood. Urine passes from the kidney to the urinary bladder via the ureter. The urinary bladder is a small muscular sac that drains through the urethra. Locate the ureters and urinary bladder.

10. Find the reproductive structures in your rat. In the male, note the penis on the outside of the body. Inside the abdominal cavity find the white, feathery Y-shaped seminal vesicles which produce seminal fluid. On the floor of the abdominal cavity locate the pair of oval testes where sperm are made.

11. In the female, locate the Y-shaped uterus. Find the two horns of the uterus which come together to form the vagina, which opens to the outside, between the anus and the urethra. Above the two tips of the uterine horns are the oviducts or fallopian tubes. Each ends at a very small, round white structure, the ovary. The fallopian tubes often coil around the ovaries, making them difficult to see. Eggs are produced in the ovaries and released into the fallopian tubes for passage to the uterus.

12. View a dissection of a rat with a sex that is different from that of your rat so that you can be familiar with both male and female organs.

3: CONSTRUCTION OF A PHYLOGRAM

PROCEDURE

1. Complete Table 1 by placing a "+" or a "0" in the blank section. If the phylum exhibits the characteristic in question, place a "+" in the blank space. If the characteristic is not present in the phylum, place a "0" in the blank.

Table 1. Summary of characteristics of animal phyla

	Characteristics			
	Distinct tissues (+ or 0)	Bilateral symmetry (+ or 0)	True coelom (+ or 0)	Deuterostome (+ or 0)
1. Porifera				
2. Cnidaria				
3. Platyhelminthes				
4. Nematoda				
5. Mollusca				
6. Annelida				
7. Arthropoda				
8. Echinodermata				
9. Chordata				

2. Compare the animal phyla to one another and indicate the number of shared "+" characteristics between each pair.

Table 2. Total number of shared characteristics

Total number of shared characteristics							
1, 2 =	2, 3 =	3, 4 =	4, 5 =	5, 6 =	6, 7 =	7, 8 =	8, 9 =
1, 3 =	2, 4 =	3, 5 =	4, 6 =	5, 7 =	6, 8 =	7, 9 =	
1, 4 =	2, 5 =	3, 6 =	4, 7 =	5, 8 =	6, 9 =		
1, 5 =	2, 6 =	3, 7 =	4, 8 =	5, 9 =			
1, 6 =	2, 7 =	3, 8 =	4, 9 =				
1, 7 =	2, 8 =	3, 9 =					
1, 8 =	2, 9 =						
1, 9 =							

3. Construct a phylogram of animal phyla. More characteristics are needed to make an accurate phylogram. Construct a phylogram with the information you have.

4. Using the information in Table 1, complete Table 2 by writing in the correct phyla.

Table 2. Summary of characteristics of animal phyla

Characteristic	Expression of characteristic	Phylum
Distinct tissues	Tissues absent	
	Tissues present	
Symmetry	Radial symmetry	
	Bilateral symmetry	
Body cavity	No body cavity	
	Acoelomate	
	Pseudocoelomate	
	Coelomate	
Fate of the blastopore	No blastopore	
	Protostome	
	Deuterostome	
Segmentation	Segmentation absent	
	Segmentation present	

288

5. The phylogram in Table 3 shows evolutionary relationships in the animal kingdom. The completed boxes on the tree are characteristics that are used to separate groups of organisms from one another. Write the correct phylum in the empty boxes.

Table 3. Phylogram of the animal kingdom

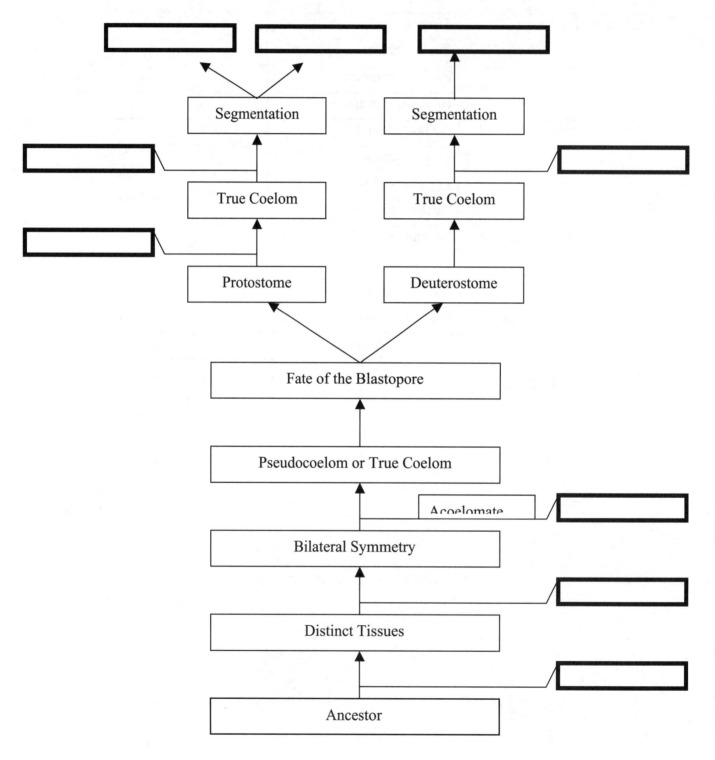

VOCABULARY

TERM	DEFINITION
Water vascular system	
Madreporite	
Aboral surface	
Spines	
Oral surface	
Mouth	
Cardiac stomach	
Tube feet	
Hepatic caecum	
Gonads	

TERM	DEFINITION
Stone canal	
Ring canal	
Radial canal	
Ampullae	
Thorax	
Abdomen	
Thymus	
Diaphragm	
Heart	
Trachea	
Bronchus	

Lungs	
Liver	
Gall bladder	
Stomach	
Intestine	
Mesentery	
Pancreas	
Spleen	
Kidneys	
Urinary bladder	
Ureter	

Exercise 20

TERM	DEFINITION
Urethra	
Penis	
Seminal vesicles	
Testes	
Uterus	
Vagina	
Fallopian tubes	
Ovary	

Exercise 20

QUESTIONS

1. The starfish and the rat share what characteristic that you have not seen before?

2. Review the characteristics used to construct the Kingdom Animalia phylogram.

Exercise 20

EXERCISE 21

DEVELOPMENT

OBJECTIVES
- Define the vocabulary terms from the questions at the end of this exercise.
- Recognize the role of differentiation and cell movement in embryo development.
- Observe the sequence of development of the starfish and identify the cleavage, morula, blastula and gastrula stages on a prepared microscope slide.
- Observe the sequence of development of the frog and identify the cleavage, blastula, gastrula and neurula stages on a prepared microscope slide.

TOPICS
1: VIDEO—"THE SEARCH FOR THE ORGANIZER"
2: STARFISH DEVELOPMENT
3: FROG DEVELOPMENT—CLEAVAGE AND BLASTULA FORMATION
4: FROG DEVELOPMENT—GASTRULATION
5: FROG DEVELOPMENT—NEURULATION
6: FATE OF THE GERM LAYERS
7: FROG DEVELOPMENT—SUMMARY

INTRODUCTION
   All sexually reproducing, multicellular organisms begin life as a single cell, a zygote, the result of a fusion between a haploid egg and a sperm cell. Yet many adult multicellular organisms consist of literally trillions of cells, each specialized to carry out a specific task. The study of fertilization and the process by which a simple zygote becomes a complex multicellular organism is referred to as embryology. Developmental biology includes the study of embryology and other developmental processes such as aging, metamorphosis and regeneration. It is one of the fastest growing fields in all of biology. Modern developmental biologists combine the classic work of embryologists and anatomists with new techniques borrowed from disciplines like molecular genetics, cellular biology and immunology. These studies have far-reaching implications for subjects as diverse as cellular communication, regeneration of injured tissue and the evolutionary relationships among animal phyla. Today's exercise serves as a brief introduction to this dynamic topic.

   Although the specific pattern of development varies from organism to organism, there are a number of fundamental processes that underlie these patterns in all multicellular organisms. The primary goal in this laboratory is to understand these processes. You will examine how developmental processes operate in starfish and frogs.

1: VIDEO: "The Search for the Organizer"

The pathway from zygote to adult, i.e., from one cell to many, will obviously involve both cell division and growth of the organism. However, the processes of cell division and growth do not by themselves explain the changes that the organism is undergoing. Cells are not just becoming more numerous. They are also becoming specialized to perform particular functions. Although all of the cells in most organisms have the same genes, only a certain number of these genes are actively expressed in each different type of cell. Over the course of development, certain genes must be "turned on" and others "turned off." This process of cell specialization, of turning genes "on" and "off," is referred to as differentiation.

Hans Spemann, Walter Vogt and Hilde Mangold were developmental biologists who worked on animal development during the early part of the twentieth century. The questions that fascinated them involved mechanisms of cellular differentiation. For example, when is it determined that one population of embryonic cells will form muscle cells and another skin cells? Is it possible to predict the future of early embryonic cells? What causes some cells to develop into certain tissues? Scientists performed experiments to answer these questions. Keep the experiments in mind as you study examples of developmental stages in lab today.

PROCEDURE

1. As you watch the video, pay close attention to the use of the following terms: blastula, blastocoel, blastopore, yolk plug, archenteron, dorsal lip of the blastopore and organizer.

2. Answer the following questions as you watch the video.

   1) As cleavage progresses in an early embryo, does the size of cells increase, decrease or remain the same?

   2) The hollow ball stage of an early embryo is called a _____.

   3) The cavity inside the hollow ball is called a _____.

   4) The dimple that forms on the blastula where cells migrate inward is the _____.

   5) The region near the yolk plug becomes the (anterior, posterior) of the embryo.

6) The embryonic _____ forms the digestive tract.

7) What part of the embryo organizes the differentiation of other cells around it?

8) Another name for the structure in the question above is the _____.

9) The parallel ridges on the upper embryo fuse to form the _____ system.

10) Hans Spemann performed an experiment to determine what happens to the two halves of a fertilized egg. Do the two halves differentiate into specific structures, for example one half into a head and the other half into a tail? Or does each half become a complete embryo?

11) Spemann divided an embryo after the blastopore was formed so that only one half had a blastopore. What happened to the half that had the blastopore? What happened to the half that did not have a blastopore?

2: STARFISH DEVELOPMENT

In this activity you will examine the early development of a marine invertebrate, the starfish. The developmental stages are easy to observe.

PROCEDURE

1. Work individually. Obtain a prepared slide of starfish development from the side bench and find the stages listed below. The slide is a mixture of different developmental stages. Sketch each stage in the space provided on the next page.

   Primary oocyte. The primary oocyte is almost perfectly spherical with a small amount of evenly distributed yolk. The large nucleus with a dark-stained nucleolus is clearly visible. Meiosis is not yet complete.

   Fertilized egg. After fertilization meiosis is completed and a polar body is produced at the animal pole. A pale fertilization membrane forms around the zygote. The fertilization membrane aids in preventing additional sperm from entering the egg.

   Early cleavage. The zygote divides by mitosis and cytokinesis to form two daughter cells that are each half the size of the zygote. Daughter cells do not increase in size between cleavage divisions so the daughter cells of the next division are smaller. The two-celled stage that is the result of the first division is easily identified. Four- and eight-celled embryos may be visible. The plane of mitotic division varies as development proceeds.

   Morula. The morula is a solid clump of sixteen to thirty-two cells.

   Blastula. As cleavage continues cells form a sphere with a cavity in the center. This fluid-filled cavity is the blastocoel. In the starfish the blastula is one cell layer thick and the cells have cilia. Even though there are many more cells, the overall size of the embryo is almost the same as the original egg cell.

   Gastrula. Imagine pushing your fist into one side of a ball to form a double-walled cup. The resulting structure is a gastrula that is formed by invagination of a portion of the blastula. The opening of the cup is the blastopore, which will become the anus of the starfish. The space created by invagination of the blastula is called the archenteron, or primitive gut. A second opening, the mouth, will complete the formation of the primitive gut. The cell layers formed by this process are called germ layers. The outer layer is the ectoderm and the inner layer is the endoderm. In the starfish two pouches form and separate from the endoderm to form the mesoderm.

2. Obtain a prepared slide of a starfish larva from the side bench and sketch the larva. The gastrula develops into a free-swimming larva that is bilaterally symmetrical. Note the gut with mouth and large stomach. The slender tube leading to the anus is the intestine. The larva swims by means of ciliated bands, and after several months, undergoes metamorphosis into a radially symmetrical adult starfish.

3. Examine the adult starfish on the side bench. The adult starfish is radially symmetrical.

PRIMARY OOCYTE	FERTILIZED EGG	MORULA
BLASTULA	GASTRULA	LARVA

Starfish development prepared slide

3: FROG DEVELOPMENT—CLEAVAGE AND BLASTULA FORMATION

Now you will study the early development of the frog. Like humans, the frog is a vertebrate animal with a bony vertebral column and a well-developed nervous system.

The first change in the newly fertilized zygote is its rapid division into two daughter cells, then four, then eight and so on. These cell divisions are typical mitotic and cytoplasmic divisions, except for the fact that there is no growth phase that separates each round of cell division. The embryo is not getting any bigger. This pattern of early, rapid cell division unaccompanied by cellular growth is referred to as cleavage. Eventually, as the cells continue dividing, a space will form in the center of the embryo. This space is the blastocoel. At the end of cleavage the embryo is referred to as a blastula.

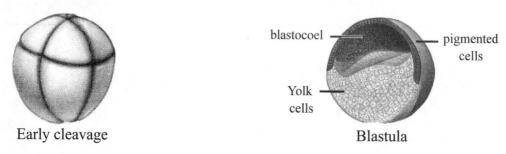

Early cleavage   Blastula

Figure 1. Frog—early cleavage and blastula

PROCEDURE

1. Using the compound microscope, examine a prepared slide of frog cleavage and sketch it in the space below.

2. Using the compound microscope, examine a prepared slide of a frog blastula and sketch it in the space below. Label the blastocoel, yolk cells and pigmented cells in your sketch.

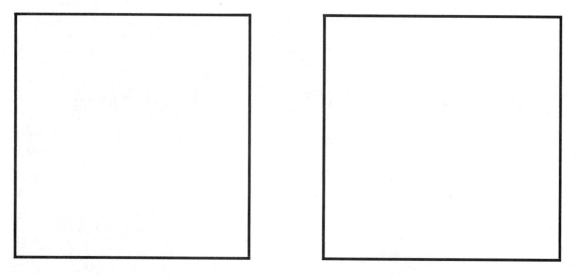

Frog early cleavage from a prepared slide   Frog blastula from a prepared slide

Exercise 21

4: FROG DEVELOPMENT—GASTRULATION

The next major change is gastrulation. An indentation or small pit on the surface of the blastula develops. This pit is called a blastopore. It represents the future anal region of the frog. The blastopore is formed as small pigmented cells on the surface of the embryo roll into the interior and migrate toward the head away from the original indentation. These migrating cells create a pocket that pushes into the blastocoel, forming a secondary cavity called the archenteron. The archenteron is an embryonic gut. As the archenteron grows, it gradually obliterates the blastocoel. Meanwhile, the smaller pigmented cells continue to spread over the larger yolk-filled cells, eventually forming a circular indentation surrounding a plug of yolk-filled cells called the yolk plug. At the completion of gastrulation there are three germ layers present: endoderm, ectoderm and mesoderm.

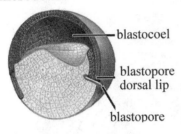

Early gastrula

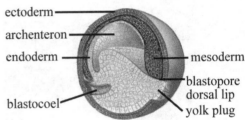

Middle gastrula

Figure 3. Frog gastrulas

PROCEDURE

1. Examine a prepared slide of a frog gastrula and sketch it in the space below. Label the blastopore dorsal lip, blastocoel, archenteron, ectoderm, endoderm, mesoderm, yolk plug, pigmented cells and yolk-filled cells.

Frog gastrula from prepared slide

5: FROG DEVELOPMENT—NEURULATION

    Neurulation is the process by which a hollow central neural tube, the future brain and spinal cord, is formed. Neurulation begins immediately following gastrulation. Ectodermal cells on the dorsal surface of the gastrula increase in height. These thickened cells form a neural plate. The neural plate is broader at the anterior end of the embryo where the brain will form. The cells at the edge of the neural plate continue to elevate, forming two neural folds running down the length of the frog from head to tail. The neural folds are separated by a midline groove called the neural groove. The neural folds continue to grow taller until they eventually move toward one another and fuse to form the neural tube. The neural tube is subsequently covered over by a layer of surface ectoderm. The notochord, formed from mesoderm, lies below the neural tube. The embryo is now referred to as a neurula.

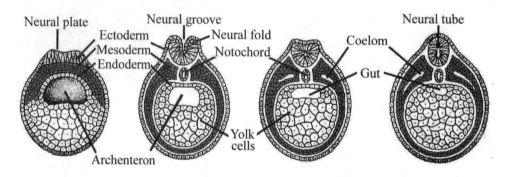

Figure 4. Neurulation in a frog embryo

PROCEDURE

1. Examine the illustrations of neurulation to help you understand what you saw in the film.

2. Examine a prepared slide of a frog neurula cross section and sketch it in the space below. Label the archenteron, ectoderm, endoderm, mesoderm and neural tube.

Frog neurula from prepared slide

6: FATE OF THE GERM LAYERS

The movement of cells from one place to another within an embryo can have profound consequences for the developmental fate of those cells. The genetically controlled movements that you are now observing transform the embryo into an organism composed of three primary germ layers: ectoderm, endoderm and mesoderm. Ectoderm is the superficial layer of pigmented cells that will come to surround the entire embryo. Endodermal cells are the large internal cells filled with yolk. Mesoderm is composed of cells that migrate from the edges of the blastopore into the interior of the embryo between the endoderm and ectoderm. These three germ layers have distinct developmental fates. The ectoderm will form the epidermis of the skin as well as the brain and spinal cord; the endoderm will form the lining of the digestive and respiratory systems; and the mesoderm will form muscle, bone and other connective tissues, gonads, kidneys and the circulatory system.

Table 1. Fate of the germ layers formed during gastrulation

	Central nervous system
	Enamel of teeth
Ectoderm	Epidermis
	Hair and nails
	Pigment producing cells
	Notochord
	Muscles
Mesoderm	Bones
	Circulatory system
	Gonads
	Kidneys
	Lining of the digestive system
Endoderm	Lining of the respiratory system
	Lining of the urinary bladder

7: SUMMARY OF FROG DEVELOPMENT

- Cleavage

  After fertilization the zygote undergoes cleavage, a series of mitotic divisions without subsequent growth. There is an increase in cell number, but no increase in overall size of the embryo.

- Blastula

  The cells formed by cleavage become oriented into a hollow ball called a blastula. The fluid-filled cavity inside is the blastocoel.

- Gastrula

  Gastrulation begins when the blastopore appears on one side of the blastula. Cells migrate inward through the blastopore to form a deep, narrow pouch called the archenteron. As the archenteron continues forming, the blastocoel is obliterated.

- Neurula

  Neurulation is the process by which a hollow central neural tube is formed. Ectodermal cells increase in height to form a neural plate. Cells at the edge of the neural plate continue to elevate, forming two neural folds running down the length of the organism from head to tail. The neural folds continue to grow taller until they eventually fuse to form the neural tube.

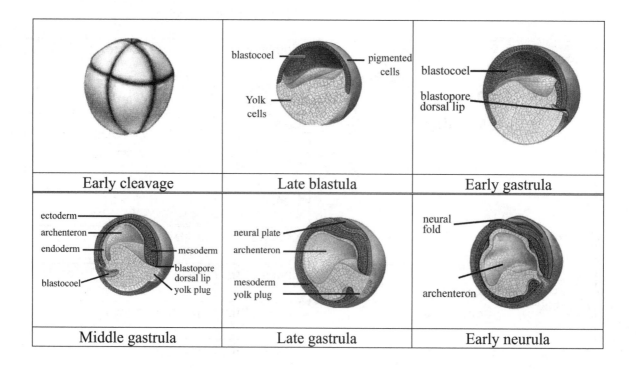

VOCABULARY

TERM	DEFINITION
Zygote	
Cleavage	
Blastula	
Blastocoel	
Blastopore	
Gastrula	
Archenteron	

TERM	DEFINITION
Yolk plug	
Dorsal lip of blastopore	
Organizer	
Ectoderm	
Mesoderm	
Endoderm	
Neurula	

Exercise 21

TERM	DEFINITION
Notochord	
Neural plate	
Neural fold	
Neural groove	
Neural tube	

QUESTIONS

1. Review the sequence of developmental stages in the starfish and the frog.

2. Review your drawings from this activity.

3. Label the drawing and identify the stage.

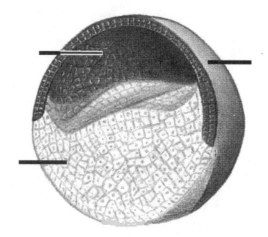

4. Label the drawing and identify the stage.

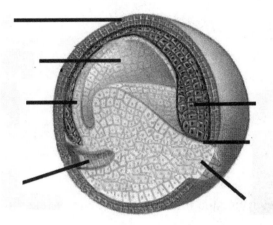

5. Label the drawing and identify the stage.

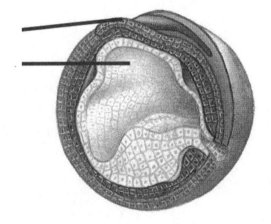

6. Label the drawing and identify the stage.

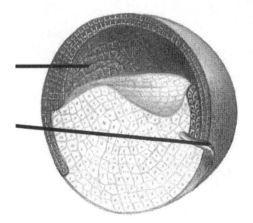

7. When do permanent differences appear in the salamander embryo?

8. How did scientists determine the future of cells that migrate through the blastopore?

9. What is the function of the dorsal lip of the blastopore? What is this area called?

EXERCISE 22

ANIMAL BEHAVIOR
PLANARIAN RESPONSE TO AN EXTERNAL STIMULUS

OBJECTIVES
- Define the vocabulary terms from the questions at the end of this exercise.
- Distinguish between kinesis and taxis.
- List and describe five types of taxis observed in planarians.
- After preliminary observations, state a hypothesis to explain planarian movement in response to an external stimulus.
- Design and perform an experiment to test the hypothesis.

TOPICS
1: DESIGN AND PERFORM AN EXPERIMENT TO TEST PLANARIAN BEHAVIOR IN RESPONSE TO AN EXTERNAL STIMULUS

INTRODUCTION
Planarians are in the flatworm phylum, Platyhelminthes. Most planarians are free-living and are common in freshwater habitats. They are also found in marine and terrestrial environments. Planarians display bilateral symmetry, meaning the right and left halves are approximately mirror images of each other. Planarians have an anterior, or front end, and a posterior, or back end. They also have a dorsal or top side and a ventral or bottom side.

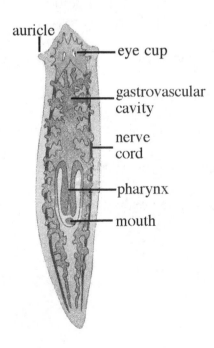

Figure 1. Anatomy of a planarian

The nervous system of planarians consists of an anterior "brain" consisting of large ganglia. Two nerve cords run from the ganglia the length of the body. Between the nerve cords are transverse nerves arranged like a ladder. In the anterior end are two eye cups containing photoreceptors that are stimulated by light. Planarians usually orient themselves to light so that the two eye cups are equally stimulated. Also in the anterior end are pits containing chemoreceptors that respond to certain chemicals. Planarians move either toward or away from concentrations of dissolved chemicals such as food. A typical planarian has a digestive space known as the gastrovascular cavity. This highly branched cavity is the site of both digestion and circulation of nutrients. The gastrovascular cavity is often referred to as the intestine or the gut. Planarians have a single opening, the mouth, found on the ventral surface. This is the site for both ingestion of food and discharge of wastes. See Figure 1.

Two typical types of movement are kinesis and taxis. Kinesis is a nondirected movement in response to environmental factors such as light, temperature, moisture or chemical cues. When an organism is in an unfavorable location, it may increase its speed, angle of movement or amount of turning. If effective, the actions will remove the organism from the adverse location. When more suitable conditions are encountered, the animal simply moves less and stays in that area. Kinesis results in random movement into better locations. Taxis, on the other hand, involves directed movements; that is, animals detect the source or differences in intensity of some factor, then they move accordingly. This movement can be toward or away from the stimulus. A positive taxis is a movement toward a stimulus; a negative taxis is a movement away from a stimulus. Taxes are frequently identified with the environmental condition to which the organism is responding. Taxes are simple responses to a stimulus, yet they are not as random as kineses because they involve movement in a definite direction. There are several planarian taxes that you can observe in the laboratory:

- Thigmotaxis – a response to touch

- Phototaxis – a response to light

- Chemotaxis – a response to a chemical stimulus

- Gravitaxis – a response to gravity

- Rheotaxis – a response to water current

1: OBSERVATION

PROCEDURE

1. Place a planarian in an observation container filled with water. Each student will observe one planarian for five minutes.

2. Note how the planarians move for five minutes and record your observations in Table 2. For example, how do the planarians orient to the light? Do they prefer deep or shallow water?

3. Record your observations in the planarian #1 boxes and the observations made by your team members in the planarian # 2-4 boxes.

Table 2. Observations of planarians before the experiment

	Minute 1	Minute 2	Minute 3	Minute 4	Minute 5
Planarian #1					
Planarian #2					
Planarian #3					
Planarian #4					

2: PROBLEM

What could cause the planarians to move the way you observed? What taxis do you want to test? You do not have to use one of the taxes already mentioned. Also, you may prefer to test a kinesis rather than a taxis.

PROCEDURE

1. Which stimulus causes the most obvious response?

2. What behavioral response do you want to test? See Table 1 for a list of possible experiments.

Table 1. Possible experiments

Taxis	Experiment
Thigmotaxis a response to touch	Place a planarian in a large container of water. After the planarian begins to swim, touch it gently with the pipette. Observe the response.
Phototaxis a response to light	Place a planarian in a large container of water. Shine a light into one end of the observation container and observe the response. Is the planarian positively or negatively phototactic?
Chemotaxis a response to a chemical stimulus	Place a planarian in a large container of water. Place a small pinch of food in one corner of a large observation container of water. Place a planarian at the other end of the container and observe the response.
Gravitaxis a response to gravity	Place a planarian in a large container of water. Take a depth reading of the location of the planarian with a ruler. Observe where the planarian is located. Is it benthic (bottom dwelling) or planktonic (free-floating)?
Rheotaxis a response to water current	Place the planarian in a container filled halfway with water. Make a gentle, circular current in the container by squeezing air out of a plastic pipette and observe which direction the planarian moves.

3: HYPOTHESIS

PROCEDURE

1. State a hypothesis to predict how the planarian will respond to the stimulus that you choose.

4: PREDICTION

PROCEDURE

1. In the space below, make a prediction based on your hypothesis.

2. State your prediction in the form "If……., then……:.."

5: EXPERIMENT

Design the experiment with the following guidelines or suggestions in mind:

- Do not choose an experimental approach that would jeopardize the planarians. Your animals <u>must</u> be alive at the completion of the experiment.

- Follow the scientific method. Be very sure you are testing only one variable. If you are in doubt, consult your lab instructor before you begin.

- Observe the response of the three planarians to the stimulus you choose until you see a pattern of movement. What is the movement? Does the movement appear to be a kinesis or a taxis?

- Record location or response of the planarians at definite timed intervals, for example every ten seconds or every minute.

- Put the planarians at different locations when you place them in the container, if appropriate to your experimental design.

- Describe the procedure in sufficient detail so that it can be repeated in the future. A good experiment must be repeatable.

- Use the grids on the following pages to determine specific location of the planarians, if applicable to your experiment. Summarize the overall movement of the planarians. Use numerical data wherever possible.

PROCEDURE
  1. Materials and Methods
     Design an experiment to test your hypothesis. Describe the procedure in detail in the space below.

Record your observations in the planarian #1 boxes and the observations made by your team members in the planarian # 2-4 boxes.

  2. Results
     State the planarian behavior and/or location at two-minute intervals in Table 3. Use numerical data wherever possible. Record the planarian response in the planarian #1 boxes. Record planarian response by your team members in the planarian # 2-4 boxes.

Table 3.   Experimental responses of planarians

	2 minutes	4 minutes	6 minutes	8 minutes	10 minutes
Planarian #1					
Planarian #2					
Planarian #3					
Planarian #4					

Exercise 22

6: CONCLUSION

Based on your experiment, draw a conclusion. Does it support your hypothesis or not? State your conclusion below.

VOCABULARY

TERM	DEFINITION
Bilateral symmetry	
Eye cup	
Photoreceptor	
Chemoreceptor	
Gastrovascular cavity	
Anterior	
Posterior	

320

TERM	DEFINITION
Ventral	
Dorsal	
Kinesis	
Taxis	
Thigmotaxis	
Phototaxis	
Chemotaxis	
Gravitaxis	
Rheotaxis	

Exercise 22

Gravitaxis	
Rheotaxis	
Benthic	
Planktonic	

QUESTIONS

1. How many variables should you test at one time? Why?

2. Why do you record movement of the planarians at timed intervals?

3. What is the advantage of having as large a population sample as possible?

4. Should planarians be started at different locations in an experimental chamber? Why?

ECOLOGY OF A ROTTING LOG

OBJECTIVES
- Define the vocabulary terms from the questions at the end of this exercise.
- Explain the trophic levels of an ecosystem.
- Construct a food chain and a food web.
- Calculate the index of diversity of a small ecosystem.
- Recognize the ecological principals underlying the construction of a pyramid of numbers and a pyramid of energy.

TOPICS
1: EXAMINATION OF A ROTTING LOG
2: CONSTRUCTION OF A FOOD CHAIN AND A FOOD WEB
3: ANALYSIS OF ECOLOGICAL NICHES. INDEX OF DIVERSITY
4: PYRAMID OF NUMBERS
5: PYRAMID OF ENERGY

INTRODUCTION

All living systems require nutrients for growth. These nutrients move from living to nonliving reservoirs and back again in a non-ending cycle. The growth, development, death and subsequent decay of a tree is one such cycle. The tree acquires inorganic nutrients necessary for its growth from the soil, air and water. Its nutrients are passed on to other organisms. These organisms can be categorized according to their positions in a food chain. The tree and other photosynthetic organisms are primary producers, the first trophic level. They capture energy from the sun to make organic molecules that all other organisms use for their nutrients. Herbivores, organisms that feed on plants, occupy the position of primary consumers and are the second trophic level. Omnivores, organisms that feed on plants and animals, are both primary and secondary consumers so may occupy either the second or third trophic levels, depending on what they are eating at the time. Since carnivores feed on other animals, they can occupy more than one trophic level depending on what they are eating.

Detritovores, or decomposers, serve an important role in breaking down organic material into inorganic nutrients that are released and made available to other organisms. Bacteria and fungi are detritovores. A detritus food chain is composed of organisms at every trophic level as they decay. The two food chains are closely linked to each other, which insures that no matter is ever lost. This rule of conservation serves as a basic principle of ecology.

As the plant's energy is made available, first to organisms on the second trophic level and then to organisms at higher trophic levels, some energy is lost at each level. Most of this energy escapes as heat. At each successive level less energy is available to organisms at that level, thus limiting their numbers. Many ecological principles can be illustrated by collecting data on a well-defined ecosystem such as the one that develops in a rotting log.

1: EXAMINATION OF A ROTTING LOG

The class will divide into teams of four. Each team will study a well-rotted log.

PROCEDURE

1. Begin with the observation and collection of fungi from the outside of the log. Look for hyphae penetrating the interior of the log. Fungi are important decomposers. Fungi indicate that decomposition is under way.

2. If termites are not available in your log, obtain one from another log. Detach the abdomen of the termite with a pair of forceps and place it on a glass slide with a drop of water. Apply a coverslip and gently squash the specimen with a cork. Observe the slide under the microscope. Several different kinds of protozoans should be evident. Sketch one or two in the space provided. Termites by themselves do not have enzymes to digest cellulose in the wood, but the protozoans have bacteria in their digestive tract that can digest cellulose. The symbiosis between the termites, protozoans and bacteria is called mutualism because all the organisms benefit from the interrelationship.

Protozoan from termite abdomen

3. Collect all the organisms you can find in your log. Identify them using the resource information on the side bench. As the animals are collected, sort them according to type and place them in covered Petri dishes. Make estimates of the numbers of each kind of organism and find their trophic level. Record your results in Table 1.

Table 1. Organisms found in a rotting log

#	Common Name	Quantity	Trophic Level
1			
2			
3			
4			
5			
6			
7			
8			
9			
10			
11			
12			
13			
14			
15			

2: CONSTRUCTION OF A FOOD CHAIN AND A FOOD WEB

A food chain is a linear arrangement of organisms that begins with the primary producer and ends with the organism occupying the highest trophic level, usually a carnivore. A food web is a group of interrelated food chains, and illustrates all feeding relationships of a community. A food web reflects the complexities of relationships in an ecosystem.

PROCEDURE
1. In the space provided below, construct a simple food chain from producer to consumer that could exist for organisms found in your log. The log itself a primary producer.

Log

2. Construct a food web including as many of the organisms that you found in your log as possible.

3. Construct a decomposer food chain for the organisms found in your log.

3: ANALYSIS OF ECOLOGICAL NICHES. INDEX OF DIVERSITY

Species diversity increases as the log decays because there are an increasing number of niches available for organisms to occupy. A niche is an organism's position or function in a community. The wood of a standing tree provides few niches for prospective inhabitants. However, a well-rotted log has many niches. Usually low diversities occur in very harsh environments or in areas just being colonized. High diversities occur in stable or more evolved communities. Ecologists use an index of diversity to compare different ecological communities.

PROCEDURE

1. An index of diversity can be calculated from the data in Table 1 to evaluate the relationship between number of species and number of organisms. Index of diversity can be calculated from the following formula where N_x equals the number of individuals of each species and N equals the total number of individuals of all species.

$$D = 1 - \left[\frac{N_1 (N_1 - 1)}{N (N-1)} + \frac{N_2 (N_2 - 1)}{N (N-1)} + \frac{N_3 (N_3 - 1)}{N (N-1)} \right]$$

2. For example, suppose we have a collection of 100 individuals composed of three species.

    species 1 = 10 individuals
    species 2 = 30 individuals
    species 3 = 60 individuals

$$D = 1 - \left[\frac{10 (9)}{100 (99)} + \frac{30 (29)}{100 (99)} + \frac{60 (59)}{100 (99)} \right]$$

$$D = 1 - 0.46 = 0.54$$

Diversity values range from 0.0 to near 1.0. A value of 0.0 indicates no diversity; numbers toward 1.0 show increasing diversity.

3. Calculate the Index of Diversity for your log using information from Table 1.

Exercise 23

4: PYRAMID OF NUMBERS

In an ecosystem there are usually fewer organisms at higher trophic levels and many more on the lower levels. Each predator usually consumes a large number of prey during its life span. The prey is almost always smaller in size than the predator and must reproduce at a faster rate than the predator in order for enough of them to survive to reproduce again.

<u>PROCEDURE</u>

1. Using the information that you have already gathered, construct a pyramid of numbers in the form of a bar graph. You will have only the right side of the pyramid. In this example, the tree is only one species.

2. Complete the graph using information from Table 1.

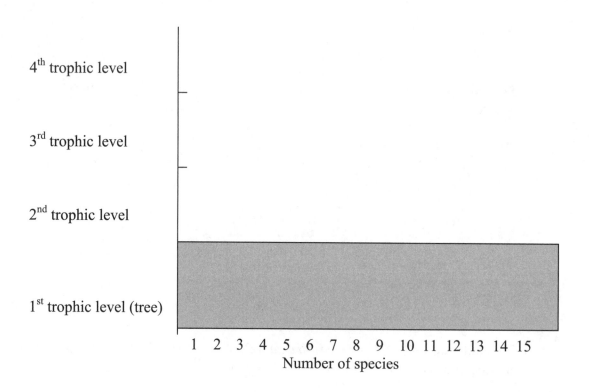

5: PYRAMID OF ENERGY

The pyramid of energy illustrates how much energy is lost at each trophic level. Frequently only 10% of the energy in a lower trophic level is stored at the next trophic level. Assume 2200 Kcal/lb are available in a log. Build an energy pyramid by calculating the calories lost at each level. How many trophic levels could your log conceivably support?

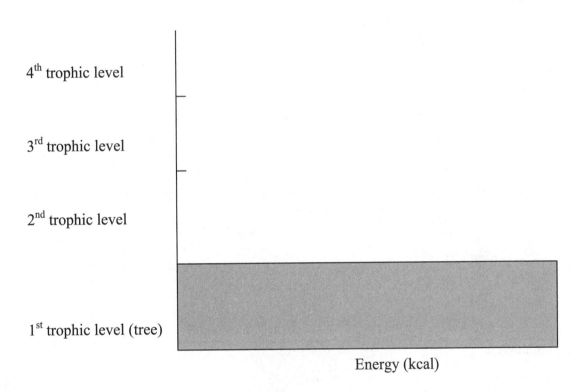

4th trophic level

3rd trophic level

2nd trophic level

1st trophic level (tree)

Energy (kcal)

The tree has 2200kcal/lb

VOCABULARY

TERM	DEFINITION
Primary producer	
Trophic level	
Primary consumer	
Herbivore	
Omnivore	
Carnivore	
Decomposer	
Symbiosis	
Mutualism	
Food chain	

TERM	DEFINITION
Food web	
Niche	

QUESTIONS

1. What is the earth's only source of energy?

2. A hawk that eats a rabbit is feeding at what trophic level?

3. Given the same amount of available energy at the first trophic level, will a short or long food chain have more energy stored in the organisms at the highest trophic level?

4. If food were scarce, which trophic level could support the greatest number of humans?

5. Calculate the index of diversity in a community where there are five different species:
    Species 1 = 50
    Species 2 = 60
    Species 3 = 30
    Species 4 = 40
    Species 5 = 20

6. When comparing the index of diversity in two communities, you find that one community has an index of diversity of 0.9 and the other has an index of diversity of 0.2. Which community is more stable?

7. A plot of grass contains 1000 kcal. How much energy would you expect would be stored in the herbivores that are supported by this plot of grass?

8. What kinds of organisms could safely be eliminated from the world without any effect on the world's environment?

9. Draw a food chain that includes the following organisms: rat snake, wheat, hawk and mice.

10. Draw a food web that includes the following organisms: wheat, mice, grasshoppers, sparrows (omnivores), spiders, one hawk and one rat snake.

APPENDIX I

Student Safety Contract

- General precautions for handling all laboratory chemicals should include minimizing exposure. Wear appropriate eye protection and protective clothing when working with hazardous chemicals. Inform your instructor if you have an accidental exposure to any laboratory chemical.

- Never eat, drink or smoke in the laboratory.

- Dispose of broken or damaged glassware in the labeled container on the side bench. Broken glassware may not be discarded in the regular trash.

- Horseplay and disorderly conduct in the laboratory are prohibited. Never perform unauthorized experiments in the laboratory.

- Clean up work areas, including all glassware, at the end of the laboratory.

- Seek information about the hazards of working in a laboratory. Review the appropriate Material Safety Data Sheet (MSDS) if you have a question concerning how to handle or dispose of a specific chemical.

- Know the location of the fire extinguisher, the emergency eyewash and the MSDS file.

- Failure to comply with these laboratory rules will result in disciplinary action.

- Students are financially responsible for breakage of prepared slides. Grades will be held until the Department of Biological and Environmental Sciences is reimbursed for breakage. Please be careful when using slides and microscopes.

- By signing this contract, I agree to follow the above rules and any additional verbal instructions that may be given.

_____

Student signature

_____

Date

_____

Lab instructor's name

APPENDIX II

LABORATORY REPORT

A laboratory report includes the following topics written in order. Use clear, simple sentences that are easily understood by the reader.

Introduction
State the reason you chose your particular experiment. State your hypothesis.

Materials and Methods
Describe your materials and methods in sufficient detail so the experiment can be repeated.

Results
Write a short summary of your results. Be sure you have numerical results. Make a graph of your results, if possible.

Discussion
Is your hypothesis supported? Explain your results. Describe other experiments you might want to do.

References

A helpful tutorial may be found online at http://labwrite.ncsu.edu/www/